Genes, Judaism,

and

Western Ethics

Genes, Judaism, and Western Ethics

Ethical Genius or God's Voice

REVISED EDITION

William I. Rosenblum

Rev. date: 03/16/2015

To order additional copies of this book, contact:
Xlibris
1-888-795-4274
www.Xlibris.com
Orders@Xlibris.com
555460

Contents

PROLOGUE

This is a revised edition of a book that describes the Judaic contribution to Western ethics. It discusses the origin of these ethics in the words of the Hebrew Scriptures, and particularly in the words and actions of the Biblical figures known as the prophets. It attempts to explain the reasons for the appearance of this ethical system on the World stage and gives special emphasis to a genetic explanation. The book was originally written at a time of great strife between Arabs and Jews in and around the State of Israel. Unfortunately this is still the case as the revised text is being prepared. None of this turmoil served as an impetus or motive for writing the book. However, in view of current events it becomes necessary to state at the very beginning of the book, that the ethical contribution of Judaism to our society is a contribution that sets a target for human behavior but does not guarantee, and unfortunately never has guaranteed, obedience to that ethical code. The book is not meant to be a brief for the superiority of Jewish behavior. Indeed the book points out the Bible's own description of great deviation by the Israelites from the code of ethics stressed by the prophets. It is beyond the author's ability to predict the extent to which the code may influence political developments in the Middle East. But whatever may happen there, or elsewhere on this dangerous planet, there can be no doubt that a yearning for peace and the interpersonal rules for obtaining it were presented to us all in the Jewish texts that form my subject.

The major change in the revised text is the acknowledgement with appropriate documentation of the debt owed by the founding fathers of Judaism to the culture of ancient Egypt and the ethical principles described in Egyptian writings. This was well known at the time the first edition was written and its omission in that text was a serious error committed by its author. Documentation of this interchange can be found in two of the cited references:

Breasted-the Dawn of Conscience ; and Pritchard- Ancient Near Eastern Texts Related to the Old Testament. The first chapter of the revision describes the interchange between ancient Egypt and nascent Judaism.

The evolution of Judaism begins about 4000 years ago and is recorded in the Hebrew Bible. As the biblical story unfolds the reader of that story is presented with the ethical principles that became enshrined in the text and are central to the religion.

This Bible consists of a series of sections. Some sections are broken up into books or subsections. The first section of the Bible is the Torah. Often mistranslated as "the Law", this word "Torah" really means the "Teachings". The Torah is also known as The Five Books of Moses. The first book of the Torah is Genesis.

The entire Hebrew Bible is also called the TANAKH. This is an acronym or made up word. Each letter of the word stands for the first letter of the Hebrew name of a major section of the Hebrew Bible, beginning with the "T" for Torah. The sequence of books in the TANAKH differs from the sequence in the Bible as arranged by Christian sources. I will use the terms Bible, Hebrew Scriptures or Hebrew Bible interchangeably to refer to the TANAKH.

The first written appearance of the new code of ethics appeared in the Torah. In the Torah there appear personages that may be called prophets. But later parts of the Hebrew Scriptures introduce us to the thoughts of later prophets. Among these prophets, Jeremiah and Isaiah are the most famous. Separate books are devoted to their teachings. In the Torah and in the books devoted to the prophets, we are told that it was God who gave the World the new ethical code. It was God who charged the Jews with promulgating the ethical revolution.

The Hebrew Bible reflects the adoption of a new World view by the ethnic group described within its books as the descendants of Abraham. Known eventually as the Jews, it was this group that was charged with the role of following, teaching and disseminating the ethical principles of the new World view. During the 1000 years preceding the onset of Christianity the Jews presented these ethical guidelines to pagan populations. These guidelines were adopted by both Christianity and Islam in the next millennium. Today, those guidelines remain the targets toward which all of Western society and much of the rest of the World aim. The introduction of these ethical principles and their persistent recitation in Jewish texts and in the teaching of Judaism from that time forward represent the true gift of the Jews to Western civilization.

In this book I will describe the ethical principles— their relation to and modification of principles found in Egyptian texts. I will present a number of Biblical characters, including Moses and the persons known as prophets. I will discuss the Bible's description of the ways in which these personages received and attempted to disseminate the new ethical message.

In these Scriptures, Moses and the prophets are said to receive their ethical messages from God. Belief in the literal truth of such descriptions is part of one's acceptance of revelation— the actual manifestation of the Divine to humans. Belief in revelation is part of the fundamentalist view of the Bible which accepts every word as literally true and as coming directly from God. What follows in this book may be of little interest to persons believing in revelation and in the literal truth of the Bible. In what follows, I take a very different position.

I believe that the preachers of the ethical messages were very special people with an extraordinary aptitude for recognizing what were the correct and the incorrect pathways for human behavior. I then consider the possible reasons for their acquiring these special social perceptions. In this book, I neither reject nor accept the existence of God. But I do state my belief that the new ethical message was not literally dictated by God to selected persons whose task was then to simply attempt to sell the message to the masses. Instead I consider several other possible explanations for the insights of Moses and the prophets.

These possibilities include the effects of drugs, or the effects of privation and isolation, or the presence of mental illness and finally simply the effects of genius that required no other aids to manifest itself in the great social principles of these persons. But, ultimately, each of these alternatives is totally dependent upon the genetic makeup of the individuals involved. In this book I present an explanation of genes and their relationship to every aspect of the human "mind" and to human behavior. And I present the case for the existence of sets of genes that led people to promote the ethical insights presented by Judaism and by the key figures represented by characters in the Bible. In addition one must account for the acceptance of monotheism and the attendant code of ethics by a small subset of Mideastern peoples who came to be called Jews. If a genetic explanation accounts for the uniqueness of the prophets, genetic characteristics may also explain the receptivity of that small audience. As the book unfolds, I will explain what I mean by a Jewish gene and I will, indeed, make the case for their existence and for their ability to influence what we call the mind.

None of these alternatives rules out the existence of some prime mover who exists behind the scenes to control the development of genes, or to direct humans toward experience that will present them with ethical understanding. The alternatives do rule out belief in the literal truth of the entire Bible. However, because I specifically deny the literal announcement of ethical principles by God to humans, I must of necessity, discuss why such stories were developed in the first place. I will take up two different possibilities: One is that the prophets and the figure or figures represented by them and by Moses, actually believed that they had heard and/or seen God or God's "glory". It is possible that they believed God really came to them when in fact they were having hallucinatory experiences under the influence of drugs, deprivation or mental illness. The second possibility is that the ethical geniuses portrayed in the Bible and/or those who came after them to codify the Biblical texts, made up these stories in order to convince a superstitious audience who certainly did believe that a God or Gods could speak and act like super-humans and could bring about events which otherwise had no natural explanation. To support this view that the prophets could have "pretended" I will cite a large amount of evidence from religious voices of later centuries.

And now, let us begin.

CHAPTER ONE

HUMANS AS A MIXTURE
OF GOOD AND EVIL

For many millennia, humans have displayed the behavioral effects of two opposing psychological traits. On the one hand we have an urge toward violence. On the other we have an urge toward collaborative, peaceful exchange with our fellow humans. The oral traditions and the writings of different cultures reflect the relative importance of these two conflicting tendencies. The Hebrew scriptures are certainly not devoid of acts of violence and acts of deceit practiced on individuals and on nations.

For example there is the story of retribution for the rape of Dinah, Jacob's only daughter. She is raped by Shechem, a Canaanite. But Shechem, while violating an unwilling virgin, is also in love with her. He asks for marriage and ultimately both his family and Jacob's family appear to agree. Jacob's sons strike a bargain with the Canaanites. The marriage will take place if all the males in the Canaanite town agree to be circumcised. The pact is made. The circumcisions take place. And then, taking advantage of the sense of security in the Canaanite town and of the just-circumcised men, still weak and in pain, two of Dinah's brothers, Simeon and Levi, murder the rapist, the rapist's father and most of the town's men.

God, itself, is not immune from performing or ordering murderous acts. Among the many examples, I will cite only two. In a fit of anger at the degeneration of human behavior, God kills all of humanity except for Noah's family. On another occasion, God instructs the Hebrew people to commit what we now call genocide.

1

God commands the Jews to wipe out the Amalekites and to leave no trace of that people's existence.

Is the violence sometimes displayed by Israelites a reflection of God's bivalent nature and the Bible's statement that we [both the good and the bad?] are "made in God's' image"? Or might we more correctly say that God itself is a construct created by humans to reflect and explain their own conflicting impulses toward both good and evil? In either case, the Text that is often violent is also filled with words urging peace and goodwill between humans. Even the story of Dinah and her brother is followed by a critical postscript. Jacob, on his death bed, presents each of his sons with a message and an inheritance. When he gets to the two murderous sons, Simeon and Levi, Jacob tells them that "their weapons are tools of lawlessness"; "Cursed be their anger so fierce"; "I will divide them in Jacob, Scatter them in Israel". So Jacob condemns violence. He is tilting toward peace. As I outline the ethical principles placed in the mouths of Biblical figures, I should like to frame those principles in terms of their emphasis on mandates that lead to a peaceful existence as opposed to mandates that require acts of violence and that encourage attitudes which condone such acts.

Creation: The Peaceful View of the Hebrew Scriptures Versus Babylonian Violence

Peace is a theme both hidden and revealed in the Biblical text. It is hidden in the description of the formation of the universe. In the Biblical description of creation there is no conflict. God creates a series of things, beginning with a light over water and over an apparent void—the as yet unmade land. God separates the light from darkness. God creates the sky to separate water below from water above. We are never explicitly told what happens to the water above the sky. Presumably it is from this that the heavens are later created. But first God creates land in the water below the sky. God creates vegetation, and the moon, the stars and the sun. And from the waters below the sky God sends forth living creatures to populate the earth and to fly in the newly made heavens. And finally God creates humans and gives to them, for food, all the seed-bearing plants and fruits. As for the rest of creation, God gives it to the humans to rule and master.

That is the first creation story in the Bible. There is no Adam or Eve. God has created simultaneously, in "God's image", a "male and female". A few pages later, in the same Book of Genesis we have a second story of the creation of woman. In this more popular version of her origins, God creates woman from Adam's rib and Adam names her Eve. Both stories are peaceful ones. In the first, God repeatedly looked down during the course of his six day effort and "saw that this was good". At the end of the sixth day, labors complete, God finds that it is "very" good. We can even imagine God smiling.

That description stands in great contrast to parallel descriptions in ancient texts;. texts from the Mesopotamian culture from whence came Abraham, the Biblical ancestor of the Jews. That culture grew between the great rivers of the Near East; the Tigris and Euphrates in the land now known as Iraq. The cultures of Assyria and Babylon also emerged in this general area. The Bible tells us that Abraham and his father Terah walked out of these lands. In the texts left to us from that culture we are told that the creation is a byproduct of violent interaction among rival gods.

In that epic story of creation, two supernatural beings bring forth others, among whom are wild children of the gods. These children are murdered by a god and the murderer is in turn murdered. A heart is taken from the last murdered god. The heart becomes the great god Marduk. Marduk is one of the ancient gods to whom humans are sacrificed. Human sacrifice was normal in the lands occupied by progenitors of the Jews during the millennium in which this Babylonian epoch was written. Marduk now becomes king of all the Gods. He is opposed by Tiamat. She is the female member of the original pair of gods that began the whole story. She is defeated in ferocious battle with Marduk. He makes, from Tiamat's body parts, both the heavens and the earth. The other gods blame one of their number for fomenting the disturbance that led to the revolt of Tiamat. This villain is slain by Marduk. Marduk then creates humankind from the slain god's blood.

Contrast this gruesome tale which makes violence the accepted center of events with the tale of the creation that we read in the Hebrew Scriptures. There we found something new. All was peaceful. There was only one God. We imagined that God smiling as He ". . . saw all that He had made and found it very good".

The Sabbath—A Peaceful Day
of Rest and Contemplation

Immediately after the description of creation something else new enters the text. On the seventh day God rests. God's rest becomes the prototype of the Sabbath. The Bible tells us not only that *we* must rest, but also that our animals and our slaves must rest. In other words the day of rest applies to an extended household consisting of all the living creatures within it. Many pagans found the entire concept ridiculous. But in the Jewish tradition the Sabbath becomes a full day of peaceful celebration, worship and study. Mimicking God, the Jews introduced the World to this weekly day of rest and contemplation.

And what about those slaves who also were to rest on the Sabbath. Should we criticize the Israelites because they held slaves? If we do consider this an ethical lapse, then we must point out that the entire ancient World embraced slavery as a way of life. Later in Jewish history, both ancient Rome and Greece depended upon slavery for economic and military security. But in the Hebrew Bible, slavery is dealt with in a way that distinguishes Judaism from surrounding cultures. Not only were the slaves to have the Sabbath day of rest, but the Bible instructs the Jews to release their slaves from servitude at the end of a sabbatical period of years. Whether this was actually done, and the manner in which it was done, are uncertain. But what is clear is the introduction of concerns for human well being that transcended the categories of the slave and the free.

God Does Not Demand Human Sacrifice

New attitudes which counter human tendencies toward acts of violence are particularly marked in the difference between the Judaic view of human sacrifice and that of the idol worshippers around them. Certainly the sacrifice, to a god, of another human and particularly of one's own child are the mark of a society that has accepted even murder as an acceptable act if it is committed out of self interest. It is easy to ignore the question of human sacrifice because it seems so remote from our own time. Perhaps it would be helpful if we remember that humans were still being sacrificed to the gods in the Americas when Columbus accidentally bumped into the New World. Such sacrifices were made by the Canaanites who shared the lands with the Jewish people. In contrast such sacrifices are absolutely forbidden by the Jewish God. Several of the prophets cried out in anguish when they observed that Jews were once again worshipping the false gods of their neighbors and participating in the rite of human sacrifice. Leviticus, the third book of the Torah specifically forbids sacrificing children to the pagan God Moluch.

The Biblical story of Abraham and Isaac stands as a representation of the absence of human sacrifice from Judaism. God asks Abraham to take Isaac, Abraham's son, to a place of sacrifice. Abraham obeys the command to take Isaac up the mountain but there is no evidence in the text that Abraham would have obeyed God. Indeed some writers have suggested that the story represents Abraham's test of God rather than God's test of Abraham's faith. Jack Miles, in his book "God, A Biography", points out that this same Abraham has argued with God about God's death sentence on the inhabitants of Sodom. Abraham also scoffs at God's prediction that he will father Sarah's child in their old age. To Miles, Abraham is not a softy who meekly follows God's instructions. Miles suggests that Abraham would not have used the knife provided to him by the angel on that mountain of potential human sacrifice. We will never know. What the story does tell us, is that in the end God refused to permit the sacrifice of Isaac. A lamb is substituted instead. Some Christian interpreters of the Hebrew Scriptures have suggested that the entire story is a prophesy of the sacrifice of Jesus, the lamb of God, as a stand in for the sins of man, personified as Abraham. But if we accept, as do many interpreters of Christian doctrine, that the Jesus of the Gospels was not only God's son, but was at the same time sojourning on earth as a human being, then God's demand for Jesus' death would be at variance with the plain meaning of the text in Genesis. I take the latter at face value. God did not—and does not—require human sacrifice.

A World without human sacrifice is certainly a kinder, less stressful, more peaceable place to live. Abolition of all murder and of theft, adultery, and envy would make it more so. Prohibitions against these antisocial tendencies are written into the Bible both in the form of God's original instructions to Noah and later in the ten commandments. But there is much more to the new ethics of the Hebrew Scriptures.

Egyptian Influence on the Ethical Code in the Hebrew Scriptures

As documented by Pritchard, ethical precepts like those presented in the Bible can be found in texts from a variety of prebiblical cultures. However the influence on Judaism of the ancient Egyptian culture is of particular interest because of the Torahs emphasis on a long period of interchange between the Egyptians and patriarchal characters who embraced or were embraced by the God that became the monotheistic focal point of the Jewish people. We are told that Abraham spent time in Egypt and the story of Joseph in Egypt is well known, even becoming the subject of a three volume work of historical fiction that gained Thomas Mann the Nobel prize for literature. For a time Josephs' monotheistic compatriots prospered in Egypt but then we have the

Biblical story of a Jewish people who were slaves in Egypt and were led out of Egypt into what we may call Palestine by Moses.

Freud in his book "Moses and Monotheism" suggests that Moses was an Egyptian and dwells at length on a fact well known to Egyptologists; namely that one pharaoh, Akhenaton, attempted an ultimately failed religious revolution, substituting one god, albeit represented by the sun, for the many gods in the Egyptian pantheon. Freud also points out that Egyptians were circumcised and suggests that this important part of the covenant between Israelites and their God must have originated in Egypt. Karen Armstrong and numerous others have reviewed the evidence that the one God of Judaism, whose name shall never be spoken and is represented by YHWH, is really the ultimate result of a fusion of two gods, one worshipped by the Jews in Egypt before the exile and one worshiped by a related tribe that joined with the exiled Jews after their flight from Egypt. Indeed some scholars suggest that two words for God in the Hebrew scriptures—Elohim and YHWH —represent the two original gods, the latter from Egypt and the former endogenous to Palestine. The scriptures treat the two names as representing a single entity.

The Torahs emphasis on a connection between Jews and Egypt is not merely myth. Rather it is supported by discoveries made in the late nineteenth and first third of the twentieth century; discoveries of Egyptian papyri, tablets and tomb inscriptions which make clear that much of what became embodied in Hebrew scripture originated in ancient Egypt. The Egyptians believed in an afterlife and appealed to the gods to allow entry into the afterlife. The appeals often took the form of writings on the interior walls of the pyramids in which the deceased was entombed. From these writings we learn what moral principles were important to the ancient Egyptians. The gods are asked to permit entry into the afterlife because the supplicant has behaved in accord with these precepts. The reader of this book will be struck by the parallel between these much more ancient statements and the ethical principles with which we are familiar from our reading of the Bible or our experience in contemporary houses of worship. Thus various individuals wrote in their tombs that they gave bread to the hungry, clothed the naked, never did evil to anyone, never took another's property by violence. In one instance a grand vizier who lived approximately 1500 years before Moses presented his son a set of moral principles to guide life.

At about the same time as the exodus another Pharaoh presented a set of moral principles. Breasted says that we know these were translated into Hebrew and found their way into the Bible. These precepts include the following: do not encroach on the land of others, do not falsify weights, do not oppress the weak, do not gossip or say evil things about others, do not lie, do not mock or laugh at the blind or physically disadvantaged.

While both Breasted and Pritchard find moral precepts not only in the words of ancient Egyptians but also in documents from other ancient cultures that preceded the crystallization of the Jewish people and their texts, there are some striking differences between the implementation of the ethical principles in Judaism and those of more ancient peoples. Before we examine these differences let us look at the ethical precepts as they appear in the Hebrew scriptures.

Biblical Ethics—Concern For Others

At the end of the Biblical period the rabbis looked at the writings bequeathed to them and found within them 613 laws. That number, 613, is derived from mystical considerations whose explication is beyond the scope of this book. Undoubtedly a lesser or greater number of laws might be found if some other numerical agenda were in play. What is important here is not the precise number but rather the relative numbers or proportion of one type of law versus another. Early in the history of rabbinical Judaism, during the first centuries of Christianity, the fathers of the Church took a hint from Paul and promulgated the doctrine that Judaism was a sterile commitment to laws rather than to love or to faith.

Paul, who began as a rabbi, should have known better. There is an abundance of "laws" that facilitate loving or at least peaceful interaction between people and encourage empathic interactions. Let us look at the 613 instructions. They can be broken down into several categories. Some have to do only with the duties of priests, or with sacrifices in the temple. The temple was finally destroyed shortly after the death of Jesus. Neither priests nor their temple exist today. Thus it is impossible to obey a substantial portion of the original 613 laws. But among the remaining rules, the largest single category, over 100 separate instructions, concern the treatment of one human by another.

This list of ethical instructions includes not only those things forbidden in the 10 commandments but also an astonishing prescription for ethical behavior first pronounced in chapter 19, verses 9-16 of Leviticus, the third book of the Torah. There the text tells us: take care of the poor; do not seek vengeance; in the court room do not treat the poor or the stranger differently from the mighty or the citizen; do not cheat in your business dealings; do not gossip; do not keep hatred in your heart; if you see that your neighbor is in danger protect him; do not lie; pay your workers promptly; do not take advantage of the disabled or lead the weak astray. And most importantly, "love thy neighbor as thyself"—the first statement of what came to be called the golden rule.

Many centuries later, about 150 year after the birth of Jesus, two of Judaism's most famous rabbis emphasized the seminal importance of this rule. Rabbi Akiva called it the major principle of the Torah. Rabbi Hillel, when he was challenged to present the essence of the Torah in the briefest possible form, chose to paraphrase the Biblical injunction by saying "What is hateful to you, do not do unto your neighbor". He then said that the rest of the Torah was a commentary on this theme and advised his listener to go and study that commentary. Surely this is a recipe for a peaceful World.

These precepts and others are reflected in the prayers recited by Jews during the services held on Yom Kippur—the day of atonement—the most solemn holiday in the Jewish calendar. In these prayer Jews ask atonement for "our" sins—the prayers say forgive us for we have sinned. Even if the individual has not been guilty of a particular transgression someone in the group must surely be guilty. The covenant between God and the Jewish people is between God and a group, not between God and an individual. God's offer of protection for the Jewish people is contingent upon the groups' behavior and therefore it is important that each individual takes notice of and asks forgiveness for not only his own trespass but that of his co-congregant. Implied in all of this is the responsibility we have to encourage-gently one hopes-correct behavior from one another.

However, the Rabbis state that God can only forgive sins between the supplicant and God itself. An example would be failure to observe the Sabbath. If the sin is an ethical infraction concerning our behavior toward another human being we must ask that individual for forgiveness. Why then, does the Jew bother to ask God to forgive transgressions that are clearly between himself and another? One traditional commentary tells us that attempts to live a blameless life are attempts to imitate the purity of God. The admission of communal guilt reflects the belief that the entire community must attempt to imitate God's goodness if God is to protect the community. The prayers that ask for forgiveness are admissions of the community's failure to reach that height. But this does not absolve the individual from seeking forgiveness from the persons we have wronged.

Now let us examine some of the differences between the implementation of the ethical principles in Judaism and their implementation in earlier cultures.

The Egyptian Way

The Jew asked for communal forgiveness and admits the sin. In contrast the Egyptian asked for entry into the afterlife by denying that the sin was ever committed! The mention of the sin tells us there was an ethical precept, but the supplicants never admit that they are sinners. Thus we have :I did not murder, I did not allow people to become hungry, I did not commit adultery, I did not cheat by putting my finger on the scale, I did not lie, I did not gossip, etc. One wonders why the supplicant believed that they could fool the gods but for this we have no answer.

Another important difference exists between the expression of ethical precepts in Jewish forgiveness prayers and those of ancient Egypt. The latter clearly ask the gods to forgive ethical breaches between people. But, as stated in the preceding section, the Talmud teaches that God will only forgive sins between man and God. The Jew who asks God to forgive sin must also ask the victim for forgiveness.

Crime and Punishment: The Hebrew Scriptures Versus The Code Of Hammurabi

But disputes will always arise, even in the most peaceable kingdom. Maintenance of peace requires a stable and respected judiciary. Judaic concepts extend to the laws applied in judicial disputes. Thus the Hebrew scriptures tell us that there must be more than one witness of a serious crime before one can bring someone to trial for that crime. They tell us that witnesses must not be tainted by conflicts of interest. They tell judges not to be influenced by the socioeconomic class of the accused or the accuser. They provide a sanctuary to which persons can flee and in which they may remain unharmed when they have inadvertently injured others. In other words there is an opportunity for one to escape from the anger of an uninformed mob or from an irrational party whom one has unintentionally harmed. Contrary to popular belief, it is Judaic law rather then Roman law, to which we can look for the origins of many principles now cherished by free peoples.

At the time of Abraham's great emigration there were not only other creation stories but also other codes of ethics in the near East. One of these, the Code of Hammurabi, is often mentioned as an example. This Code is often presented to show that Judaism was not unique in providing ethical guidance to the pagan World. In fact, however, the Code of Hammurabi illustrates ethical principles that differ quite markedly from Judaic ideas of morality. With respect to punishment in particular, the code can serve, along with the violent story of creation described earlier, as an example of the extreme difference between the culture that emerged in Judaism and the culture from which Judaism emerged. In his code, King Hammurabi tells us that Marduk—the God who demands human sacrifice—has ordered him to protect the weak from the strong and to enlighten the land, thereby furthering the well being of mankind. But unlike the Biblical code, Hammurabi's code is, in large measure, a document that values people, not for themselves, but in accordance with their place in the socio-economic system. For example if one strikes a pregnant woman causing her to lose her child, the striker must pay five shekels to a free woman but only two shekels to a servant. If one injures a slave, the penalty— paid to the owner—is only half the penalty for the same injury done to a free man. How different from the dictum in the Bible's book of Exodus where one is told that the murder of a slave is no less a crime than the murder of any other human and that the penalty for both is death. Indeed the rabbis, in Roman times, tell us that one who kills a single person murders the whole world while one who saves a single person, saves the whole world.

The Bible does demand the death penalty for nine crimes: murder, kidnapping, incest and adultery, blasphemy, idolatry, desecrating the Sabbath, witchcraft, and dishonoring one's parents. To the modern reader this is harsh doctrine. But it is misleading. While rabbis could not change the Bible, they ruled in ways that made it virtually impossible to carry out the death penalty for crimes other than murder. Nevertheless, if capital punishment is a feature of Judaic ethics, in the millennium following the appearance of the "new morality", then we must ask whether the Judaic message was any less violent than that of the culture from which Abraham fled. In fact, Hammurabi's code is a much more violent one as befits the code of a king who is brought to power by a god that demands human sacrifice.

Like the Bible, the code of Hammurabi demands the death penalty for adultery and for certain forms of incest. Strangely it does not mention murder or set a penalty for it. But perhaps this is mentioned in one of the thirty-three laws on stone tablets so far missing from among those surviving tablets on which the code has been found. Other crimes that are presented as deserving the death penalty are: bearing false witness, theft, harboring runaway slaves, breaking into a house for the purpose of stealing [even if there is no theft]. If a jailer mistreats a prisoner and the prisoner dies then the jailer's son may be executed in compensation. In short the death penalty is extended to crimes other than those considered by the Bible to be worthy of capital punishment. Of particular note is Hammurabi's insistence upon the death penalty for crimes against property. But more striking is the element of cruelty in the punishments recommended for lesser crimes in the code of Hammurabi.

Let us consider the case of a wet nurse who is hired to suckle a child. The child dies and the wet nurse, without asking permission of her employer, suckles another baby. This is a crime—though why it is we do not know. But that is not what concerns us. Rather we are concerned with the punishment of the crime: the wet nurse is to have her breasts cut off! Or let us consider the case of the child who disavows adoptive parents, though they have rescued him from the dishonor of life with a disreputable birth parent. What is the punishment for the ungrateful words of so callous a youth? Why naturally, his tongue is to be cut out! From these and similar laws we get the picture of a doctrine which in a bizarre way makes the punishment fit—through an exercise of analogy—the crime. Bad speech—off with the tongue; bad nursing—off with the breast.

We should contrast this with the general statement in Exodus which tells us to take an eye for an eye and a tooth for a tooth. This was not a doctrine to be taken literally. The people were not to punish the offending organ. Rather they were to be sure that the punishment was proportional to the crime; that the punishment did not exceed the crime. Not two eyes when only one was injured; not death when the crime was theft rather than murder.

The idea of proportionality is the one that has been adopted by Western society. In this regard it is of interest to note that Western society has not, as a matter of law, adopted the idea espoused by Jesus. Jesus tells us to turn the other cheek rather than to punish. If an article of clothing is stolen, we are to give the thief even more of our clothing—the shirt off our back. Perhaps this was a recognition that the thief may be more needy than we. But in Western society, while we may consider mitigating circumstance—as did the rabbis from among whom Jesus sprung—we nevertheless punish proportionately to the crime, rather than reward the criminal.

The Nature of Leadership

The nature of political leadership is also important in a peaceful and therefore stable society. Moses is overwhelmed by his responsibilities. He finds himself in the position of an absolute dictator trying to bring God's order to Its people. Moses father-in-law, Jethro, is not a Jew, but he is wise man. He advises Moses to establish a hierarchy of leaders within the community, each having a subset of the people under their direction. Each cadre will report to Moses at the top of the hierarchical pyramid. Moses takes his father-in-law's advice. Similar systems are in place throughout the World today.

Leadership at the top changes with the changing generations, but the Jews had no absolute leader for several centuries. Ultimately, the scriptures tell us that they demanded a king; they wanted to be like other nations. Here, certainly, we see an illustration of cultural assimilation. This appears to contradict my proposal that the ancient Jews had, or were led by those who had, a substantial difference in World view from that in the surrounding cultures. But no new culture can ever break completely with what surrounds it.

And, even here, the writings that describe the event tell a unique story. Saul, the first king, does not want to be king. He actually hides in an attempt to avoid the responsibility. And the people are cautioned by those whose words come down to us in the Biblical book named after Samuel: if you must have a king, be careful; the king should not be one who enriches himself at the expense of the people. And I would add that a well ruled nation is more likely to be a peaceful one.

All of these examples show the distinctive World view embodied in the Scriptures; writings that I believe paraphrase the teachings of those ethical geniuses upon whose thoughts I am reflecting. When I use the word "peace" to tie these examples together, some readers may conclude that I am engaging in unnecessary poetic license. However the plea for peace does become an explicit portion of the Judaic World view in the writings of the Prophet Isaiah. It is he who prophesies the time when swords will be broken into plowshares and the lion shall lie down with the lamb.

Peace As The Goal Of Nations

Isaiah's prophecy appears to concern the nature of the World at or close to the coming of a messiah; a time close to the end of days and the disappearance of the universe and its inhabitants in their present form. But the Jewish view became one of striving to achieve that goal in the World in which we dwell. This point of view is made explicit by Jewish mystics writing in the millennium after Isaiah. The Jewish goal, they said was to restore the fractured state of the World. In the fanciful imagery of the mystics the Godhead itself was contained, in all its manifestations, in rays of light emanating to earth and captured in containers that were inadvertently broken during creation. In this way evil entered the World and other manifestations of the Godhead were also separated. Thus, for example, the feminine portion of God's spirit was separated from the Godhead. Our mission on earth is to put together the broken shards of the fractured vessels and restore the unity. To "repair the World" is to perform the good acts which will in turn restore the unity of the Godhead and bring about a state of grace that brings about the coming of the messiah. But there is an old Jewish aphorism which tells us that we will recognize the messiah when we no longer need one. The universal application of ethical principals introduced by the prophets who claimed to speak with the words of the Lord would bring about the messianic age and the peace prophesied by Isaiah even without the appearance of a messiah who seems to speak for God.

Isaiah says that Jews will be a light unto the nations. In the terms of the mystics this means that Jews must engage themselves in the colossal task of putting together the broken vessels. When the task is accomplished, Isaiah's time of peace will be at hand—swords will have been turned into plowshares and lions will lie down with cattle. So Isaiah's ideal state—a time of universal peace—is not one that will magically arrive as the end of days approaches. Rather that ideal is one toward which all humans must work, exerting enormous effort, during each of our allotted times on Earth.

Was there ever before the idea that peace is the goal toward which all human effort must aim? Was it not until the twentieth century and president Woodrow Wilson that we again heard about the end of all wars? Was it not until the advent of the League of Nations and then the United Nations that we see serious political effort—however flawed it may be—to make peace the foremost goal of nations? This peace of Isaiah is not the peace that is achieved by conquest and subjugation—the Pax Romana. It is not the peace that comes from fear—the Pax Atomica. It is peace for its own sake; peace because that is the way things should be.

CHAPTER TWO

THE PROPHETS
AND THEIR MESSAGE—
WHAT, WHY, AND WHEN?

In the preceding chapter I have reviewed many of the ethical
principles embodied in Hebrew scriptures. As noted these
principles were in whole or part present in more ancient cultures
and appear particularly to have been imported into Israelite
culture from the Egyptians. However, also noted were significant
modifications of principles, in particular the ways in which the
Israelites took responsibility for the adaptation of the principles
and for their violation. The code of ethics became a code ordered
by a monotheistic deity and their adaptation was as necessary as
the worship of that deity. Indeed one goal of obedience to all of the
tribal laws was to bring the tribe member closer to the perfection
of the deity itself. In earlier cultures there surely must have been
special individuals with a highly developed ethical sense who
introduced the ethical principles and encouraged obedience to
them just as the Pharaoh Akhenaton introduced monotheism to the
Egyptians. Similarly there must have arisen among the Israelites
similar leaders whom one might call ethically gifted and with
charismatic gifts that enabled them to influence an entire nascent
nation. Moses appears to have been the first to be described in
detail in the Hebrew scriptures. He has been called a prophet.
Others were to follow.

What Did The Prophets Do?

Today the term "prophet" connotes someone who predicts the future. But the principal mission of the Biblical prophets was something else. They were given the task of exhorting the people to obey an ethical code that separated them from their neighbors. When they did predict the future, the prophets generally described how the people would be punished if they disobeyed the new code of ethics. The punishment was to be a collective one. It would be meted out to a new Nation. That Nation was to arise from the audience to which the earliest prophets preached. As that Nation unfolded, the audience became a "people"; the Jewish people. To these people a task was given. They were to serve, through their behavior, as a beacon for humanity. This beacon was meant to lead the world to follow the ethical principles introduced by the prophetic voices represented by the Biblical leaders. The collective punishment meted out to the Northern Kingdom of Israel and then to the surviving Kingdom of Judah represented payment for the failure of the people to lead in this way. But, the prophets also predicted redemption when the people ceased to stray from the new message.

When Did The Prophetic Message Originate?

With the 10 commandments Moses presented the Israelites with an initial truncated set of ethical principles. Prior to the appearance of Moses in the Biblical book of Exodus, the first book, Genesis, records under Noah's name, some of the principles included in the 10 commandments. The more extensive and nuanced code that I described above is set down especially in the Books of the Bible called Leviticus and Deuteronomy. Tradition not only ascribes all these words to God but tells us that God gave them to Moses who then passed them on.

We cannot know precisely when the new message was first preached. Presumably the message was transferred from Egyptians to disciples whose collective influence has been subsumed under the name of Moses.

Moses is sometimes called a prophet, especially by persons who are not Jewish. But if we define prophesy as a prediction of future events then Moses made few if any prophesies. In Exodus it is God who tells Moses what will happen. God also reiterates a covenant with the Jewish people. This is in the form of a contract which demands that they not only accept the Lord but also obey His commandments. God's promise to take care of His obedient people is not really a prophecy. It is simply part of the contract. God's part of the bargain is repeated later in the Bible by later prophets. Then the promise of prosperity under God's wing might be considered as a prediction of the future by these prophets. But this is a contingent prophecy—to be fulfilled only if the people obey God's law. God's own prophecy is better illustrated by God's description of future events such as the entry of the Israelites, but not of Moses himself, into Canaan, the promised land.

Moses and the prophets called the people to the One God. That call may have been heard both by the Hebrews and by those who were not ethnically related to them. "God-lovers" may have remained uncircumcised and may not have obeyed many of the Jewish laws but may nevertheless have found the idea of the One God and many of His teachings appealing. Such people certainly existed among the otherwise pagan populations many centuries later. For example they are documented in Alexandria, Egypt during the time of the famous Jewish philosopher Philo. Moreover there were additions to Judaic thought from the surrounding non-Jewish communities. Indeed the Bible makes clear that some persons who were not ethnically "Jewish", persons whose origins were outside the 12 tribes, could still embrace the One God as their own. It is ironic that the clearest example of this is one in which the believer challenges this God, demanding the explanation for otherwise inexplicable destitution. That believer is Job, a non- Israelite from the land of Uz. The Book of Job tells us that Job was obedient to God's laws. But it does not tell us that he actually converted to Judaism or when the conversion of Job or of his forefathers to either "God worship" or to Judaism may have occurred. Scholars of course attempt to determine, from linguistic clues, when the story may have been written. But that is not at all the same as determining when the events are supposed to have taken place.

Perhaps the One God was attractive to some pagans because the Judaic message clearly stated that they were included among His concerns. That is the message of the Book of Jonah, who, like Jeremiah, is a prophet who doesn't want to be one. He is ordered by God to go to Nineveh and preach God's eternal message. Nineveh was the capital of the Assyrian empire—it was a nest of Pagans. Jonah is afraid of the citizens of Nineveh. He flees God. But his flight is not a success. Instead he is swallowed by a whale. His fate is sealed until he repents. Then he is regurgitated. He will go forth to carry out God's work. Traditional interpretations of the story tell us that by making Jonah the messenger and by commanding him to deliver the message to all of Nineveh, the text tells us that the Jewish God is everyone's God and the Jewish message is the ethical message to everyone.

Not All Prophets Were Israelites

There may even have been pagan prophets, not mentioned in the Bible. They may represent false prophets; one's whose words are lost to history because they were incorrect. Or, even if correct, their messages may have fallen on deaf ears and been lost to history because their listeners were unprepared to heed the new message. However, at least one pagan prophet did have his words recorded in the Bible. That pagan prophet was Balaam, a Canaanite who spoke to his fellow Canaanites.

Balaam is a prophet for the worshipers of Baal. He is asked by his king to prophecy about the events that will transpire in the war with the Israelites. Balaam investigates the enemy and concludes that they are a great and virtuous people who will prevail with the aid of their God. This story appears in the Bible along with a long poem by Balaam, in praise of the Jews. A portion of Balaam's words is actually used at the beginning of many traditional Jewish services: "How fair are your tents, O Jacob; your dwellings, O Israel." Balaam's king was, of course, infuriated by Balaam's prophecy. Nevertheless, and in spite of Balaam's benign view of the Jews and his prophecy of Jewish victory over the tribes of Baal, the rabbis castigate Balaam. This seems irrational to me. But perhaps the rabbis could not embrace a prophet who remained with the enemy in spite of his favorable view of the Jews.

Prophecy And Ethics During The Period When Israelite Kingdoms Were In Decline

Balaam spoke in admiration of the Jews during the period of their ascendancy. But for the most part the prophets spoke in the period of Israelite decline as a political power on the Middle Eastern stage. According to the Bible the Israelites were united under King David and his son Solomon. But the competition between Solomon's sons leads to the division of the Davidic Kingdom into two pieces, the Northern Kingdom of Israel and the Southern Kingdom of Judah. Almost immediately the Northern Kingdom is threatened by Assyria. Later, after the Assyrian victory, Judah is conquered by the Babylonians.

The Jewish prophets of the period accurately foretold these events. But the trajectory of decline was not always a straight line. Thus Isaiah tells the King of Judah that the Assyrians will fall back and will not capture Jerusalem. The prophecy is correct. Later, it is to the Babylonians that Jerusalem will fall. This is the prophecy of the prophet Jeremiah, who tells his King that there is no point in trying to prevent the Babylonian capture of Jerusalem and the exile of most of Judah to Babylon.

Isaiah and Jeremiah prophesied in the middle of the millennium before the birth of Jesus. By that time the Northern Kingdom, containing 10 of the 12 tribes or clans established during Moses' time, had disappeared into the Assyrian empire and only the Southern Kingdom of Judah survived. But earlier, when Israel flourished, its pagan neighbors paid it far more attention than they did to the much smaller Kingdom of Judah.

For part of those earlier days, when the Kingdom of Israel was strong, it was ruled by King Ahab. We know from the writings of surrounding cultures that Ahab was a historical figure. His army and especially his charioteers were much feared. Indeed when Ahab was finally conquered by the Assyrians, the conqueror kept that unit of Ahab's army intact to serve as a lethal weapon in the conquering army.

Many people will recognize the name of Ahab's wife and Queen more readily than the name of Ahab himself. That Queen was Jezebel. She is the epitome of the immoral. Elijah was the principal prophet during Ahab's reign. Jezebel is vigorously criticized by Elijah. Jezebel orders that Elijah be found and killed. He escapes and prophesies a horrible death for Jezebel. Elijah also describes the unraveling of Ahab's empire and the death of his relatives. All the prophecies come true.

What was really the "sin" of Ahab and Jezebel? The text tells us that Ahab and Jezebel abandoned the one true God. The Bible uses that desertion as a synonym for immorality. But it is clear from the words of prophets like Amos or Isaiah, that mere ritual observance of the monotheistic religion was not the principal requirement of that God. Rather it was adherence to the ethical code involving one's interaction with the surrounding World, both Jewish and pagan, that was the most important requirement. Ahab and Jezebel were immoral in the sense we all would recognize today. That immorality was the reason for their punishment. Elijah rails against the immorality. So do the other prophets during the millennium preceding the birth of Jesus. A real appreciation of the prophetic mission can only be gained by reading those Books of the Bible that are given over to the words of each individual prophet. The following few paragraphs give only the flavor of these texts.

Amos noted a number of moral lapses. Among these was the increasing size of the gap between rich and poor and the failure of the rich to obey precepts that would better insure the well being of their less fortunate fellow citizens. Jeremiah noted, among other sins, a return to the practice of human sacrifice. Ezekial was also outraged because human sacrifice, and specifically child sacrifice, had continued among Israelites attempting to placate false gods. So, in spite of God's ultimate rejection of human sacrifice in the story of Abraham and Isaac, the complaints of the prophets indicate that such sacrifice was still a common practice among the pagan tribes and among the lapsed Israelites.

In addition, Ezekial reported that God complained of a general rejection of His laws. Ezekial begins the list with an outcry against idol worship. He castigates false prophets and those who engage in magic. God is outraged at fornication committed with those of other cultures. God contrasts sin with what is right. Right behavior includes rejection of adultery and incest. Ezekial also teaches that good people oppress no one. They care for the widow and the orphan. They feed and clothe the poor. They treat father and mother with respect. They do not rob. When a debt is paid good people promptly return the security of the debtor.

These precepts are identical to those found in the Five Books of Moses. Thus a reiteration of the ethical teaching embodied in the Torah is embedded in the broad sweep of lyric or poetic texts that characterize the Hebrew Scriptures devoted to the prophets. While deserting the Lord, worshipping other Gods and participating in the religious practices of other peoples, the Jews were at the same time abandoning the ethical underpinnings of the religion, the ethical rules brought to the people by the Lord.

The prophets attempted to frighten the people into returning to the Lord. This attempt means that the bulk of some of the texts is devoted to apocalyptic visions of death and destruction levied against sinning Israel or Judah, and against other nations who fail to adopt the ways of the Lord. These other nations include Assyria and Babylon who in the shorter term will be triumphant.

But the prophets must not only present the punishment, they must also hold out hope. Without hope there is no point in returning to the ethical path. To the later prophets the ultimate punishment was the destruction of the temple in Jerusalem and the conquest of that city by the Babylonians. Hope is generally represented by extensive descriptions of an ingathering of the peoples to Israel and of the restoration of Jerusalem and the temple. The prophets Haggai and Zehariah tell us that their words were pronounced during the exile into Babylon which followed the temple's destruction. These prophets announce the imminent return to Judah. The latter event actually occurs after the Persian conquest of Babylon. The promise is also made of some more elevated salvation in or after what has been interpreted as messianic times. But that promise is, in fact, a minor thread in the prophetic tapestries, albeit a thread seized upon by the founders of Christianity.

The Need To Entwine the Belief In God With The Ethical Code

The prophecies describe retribution for the rejection of both the Lord and the Lord's instructions. The Lord and the Lord's instructions cannot exist one without the other because it is the instructions that define the difference between this God and all the others. Therefore, acceptance of the Jewish God can be viewed as a metaphor for the acceptance of the ethics of Judaism. However there are two sorts of religious edicts embedded in the Hebrew Scriptures. One is the set of ethical injunctions I have been talking about. The other consists of religious ritual; rules such as those for prayer ; rules that tell us what we can and cannot eat; rules that forbid the wearing of certain fabrics; rules that dictate our holidays.

After the Bible was written, rabbis extrapolated from these and other rules in the Text to create rules for the times in which they lived. For example, work is forbidden on the Sabbath. But what constitutes work? When the automobile was invented, the rabbis had to decide whether driving was work. The rabbinical definition of work is based on the section of the Bible that describes the work performed in building Solomon's temple. Kindling a fire is included in this list of work. The rabbis decided to forbid driving an automobile on the Sabbath because they felt that a spark was kindled in the internal combustion engine. But, as Micah says, obedience to religious rules like this one can only be meaningful if, when performing them, one remembers also, those edicts that are focused upon social concerns.

Within the river of Judaism there runs a postbiblical current that interprets ritual observance as observance which sweeps the practitioner closer to God's perfection. Within that Perfect Model there lies the ethical perfection embodied in the broader stream of prophetic teaching.

Thus Judaism is not, as the apostle Paul implied, a religion of rules about our relationship to things and objects. It is not a religion whose practitioners are doomed to religious failure because of the impossibility of following all the regulations. Judaism does not set one regulation above another; the ritual is not more important than ethical behavior. Indeed the former is without religious meaning in the absence of the latter. This is made very clear in the words of Isaiah that are part of every Jewish service on the Day of Atonement [Yom Kippur] : This is not the fast I asked for—a fast devoid of any concern for the welfare of the people—. Isaiah is telling the Jewish people that obeying God's law about ritual is only meaningful if it reminds one of the ethical injunctions of the religion. Thus both God and ritual can be viewed as metaphors for ethics. They both must represent the principles we must remind ourselves of as we contemplate either the concept of God or perform rituals that we have been told are dictated by God.

The prophets spend much time castigating the Israelites for rejecting these metaphors and hence for forgetting what they stand for. I will suggest that the prophets used these metaphors because they felt that this was the only way to connect with the people. In this book I need not offer an opinion concerning the prophets' belief in God. But I will suggest that, whether believing or not, the prophets certainly knew that God did not speak in a human voice. This assertion will be fully dealt with in a later chapter. If the assertion is correct then one may also suggest that the prophets put their message in God's mouth in order to influence their audience.

The organization of the five Books of Moses reflects this use of God's name to reinforce the ethical imperative. The narrative in the Books of Moses is often interrupted by ethical pronouncements and rules for behavior. Why not place the entire narrative in one place and the rules in another? The easiest answer is one which says that the chosen presentation is one which repeatedly reminds the audience that the ethical principles are coming from a God who also holds the future of the people and their descendants in His hands. To place all the rules in one place and the narrative in another would be to leave room in the text for only a single point of intersection linking the rules of behavior to God's place in the history of the Jewish people. It is not necessary to believe in God in order to understand the powerful effect of the actual arrangement of the text on an audience who certainly did believe in the supernatural. That audience is repeatedly reminded, during breaks in the recitation of the law, that obedience to that law is God's wish and that obedience or disobedience will determine not only their fate but the fate of their children and all the generations to follow. I was once asked by a minister to explain why Judaism asks for ethical behavior without—at least in its beginnings and even later in its lack of emphasis -the promise of heaven. Surely the powerful answer is Gods promise to guarantee the future of all our successive generations, if only we obey the law.

We cannot know whether the prophets are fictitious characters or whether the writings ascribed to any one of them are really the words of several persons rather than one. We might assume, but do not necessarily have to assume that the writings are not that of real life prophetic figures. But for now, let us assume that they were. Let us look then at the ways in which the prophets tell us that they received their messages.

CHAPTER THREE

GOD'S VOICE AND IMAGE

The prophetic Books of the Bible present a diverse picture of the prophet's experiences. Hosea, an early prophet simply tells us that the word of the Lord came to him. But the next paragraph of the text tells us that the Lord "spoke". Is this merely a metaphor or are we to believe that Hosea heard a human voice? The text is silent about this. There is no description of how God's voice sounded. No visions are reported; no fantastic and supernatural images. Rather the prophet simply recounts in poetic form the straying of the Northern Kingdom of Israel from the way of the Lord. Hosea predicts the demise of the Northern kingdom but at the same time offers people an opportunity to return to the fold and be forgiven, before it is too late.

Amos, at first reports no visions and hears no voices. When he tells us that "the Lord said" we are free to assume either that God spoke in a human tongue or that Amos is simply inspired to relay messages that he feels rather than actually hears. But then, in the middle of the Book containing his prophecies Amos announces "That is what my Lord showed me". Here begins a series of stories in which God both speaks to Amos and shows him visions. The visions are not fantastic, but rather mundane: a plumb line ; a basket of figs. From these the Lord extracts a hidden and poetic meaning which he explains to Amos. In the end the message is much like the message to Hosea. God tells Amos of the extinction of the Northern Kingdom but the preservation of some portion of the House of Jacob and ultimately the planting of the people Israel upon the soil which God had given them.

We are also told that the word of the Lord came to Zephaniah, Haggai, Zechariah and Malachi, but we are not told how. Obadiah,

Micah, Nahum, Habakkuk and Isaiah received what the translators of the text call "visions". But there is nothing in these accounts to suggest that these prophets believed that they had received an image of God. The so-called "visions" are merely reported as the word of God.

Jeremiah Is Touched By The Lord

Jeremiah tells us that the Lord put out His hand and touched the mouth of the prophet, saying "Herewith I put my words into your mouth". After this the Lord comes repeatedly to Jeremiah and speaks to him. Jeremiah not only hears God's voice; he is also presented with visions in which future events or allegorical representations of them are placed before his eyes. But Jeremiah never tells us that he sees an image of the Lord.

In fact, none of the prophets mentioned thus far in this chapter see an image of God. Perhaps this omission reflects the belief of later theologians that humans cannot really know what God is. Even Moses never sees God's face. God is an enigma. As He explains to Moses, "I am what I am".

Images Of The Lord

However Isaiah, who prophesied immediately before Jeremiah, does present us with fantastic images of the Lord. And it is quite a Lord! Isaiah's vision includes a temple and the Lord's skirts filled the entire building. Isaiah never quite sees the Lord's face, but he does see figures that surround the Lord. They are called angels and each has six wings. The angels purify Isaiah's lips by touching them with a hot coal. Of course his lips show no evidence of a burn. Isaiah with his lips now purified, volunteers to serve the Lord.

The bizarre aspects of Isaiah's vision are far exceeded by the visions in the Book of Ezekial. Ezekial prophesied during the period of Judah's exile in Babylon. He has vivid visions in which he sees a throne on which sits an image encircled by a radiance that Ezekial interprets to be the glory of God. In the center of the radiance there is a "gleam" that we may surmise represents the Lord. This gleam is accompanied by four figures. Each has four faces and four wings with human hands below them. On each of the figures one face was human, while the other faces were that of lion, ox and eagle. The details of Ezekial's visions go on and on. He describes something called the "semblance of the Lord". This image has something like "loins" which gleamed like a fire encased in a frame. Below them was something that looked like fire and above them was a radiance like a rainbow. The Lord speaks to Ezekial, giving him a prophetic mission. The Spirit lifts him up, rushes him along and sets him down again amidst the people where he remains in a daze for seven days. God then speaks to him again saying that Ezekial is now the Lord's watchman among the Israelites and is to take the Lord's warnings to them.

Thus we see in these accounts of the prophets' experiences a group of supernatural events ranging from hearing voices to seeing complex and fantastic visions. In the next chapter we will discuss several possible explanations for these events.

CHAPTER FOUR

ILLUSION OR REALITY?

The preceding chapter described the ways in which the prophets tell us that they received their messages. How are we to account for these supernatural events? The first explanation would simply accept the accounts as being the literal truth— miracles. This would be a view of the Bible which says that everything in it is literally true. As explained in the introduction to this book, the author does not accept this fundamentalist view. Instead this book directs itself to those who would like to explore alternative explanations for the accounts of the prophets. Indeed, if the prophets were simply passive vessels selected by God to parrot God's message, then we would give no credit to human genius expounding and demanding fulfillment of the ethical precepts expressed in the Biblical text. It is my desire to give credit to the ethical geniuses whose ideas are presented through the voice of Moses and the prophets. Consequently I present the reader with alternatives to the fundamentalist view of the texts. In this analysis it is permissible but not necessary to believe in the historical existence of any particular Biblical prophet. It is only required that we acknowledge that their words were uttered by real persons and ultimately accepted in principle by the people who became known as Jews.

Dreams

It is possible that one or more of the characters known as prophets, existed, and dreamed the events portrayed in the Biblical text. They may or may not have believed that the dreams were God-given. The latter possibility is discussed in a later chapter.

But, in any event, they reported the dreams as if the events they contained had really occurred. I know of no evidence that would either support or deny this suggestion.

Events Induced By Attempts To Reach An Enlightened State

On the other hand, texts exist which suggest that the experiences reported by the prophets may have come from attempts to reach what some practitioners of meditation have called an enlightened state. We have no way of knowing whether or not the prophets simply meditated to obtain their messages. But there are texts which support the idea that prophets may have used either deprivation or drugs to obtain enlightenment or to heighten their emotional state as a means of providing the motivation necessary to persevere with their message in the face of oppressive opposition. Such opposition is especially clear in the cases of Elijah and of Jeremiah. The Biblical narrative tells us they were hunted down by their respective kings and their lives were constantly threatened by the establishment. If the prophets used self induced states of ecstasy to receive their messages we cannot know whether or not they really believed that God was present and speaking during the event. A later chapter considers the possibility that they did not believe that God really appeared and that they did not believe the events they experienced really happened. But for the present I merely wish to explore the evidence that the voices and events were self- induced. The Biblical accounts themselves provide no such evidence. However a model for such behavior is provided in at least one other account of a prophet's life.

Ezra's Deprivation In The Desert And The Eating Of The Flowers

This account is provided by the man calling himself Ezra, in the book most often known as Second Esdras. Second Esdras is an apocryphal book. This means that the founders of rabbinical Judaism and many Christian theologians believed that, unlike the accepted Books of the Bible, Esdras came neither directly from God nor through divine inspiration of its human author. For that reason it never became part of the Jewish canon enshrined in the Hebrew Scriptures. A number of apocryphal books were, however, included with the first translation of these Scriptures into the common Latin of the people within the old Roman Empire. This translation into the "vulgate" was made by Jerome early in the Christian period and remains part of the printed Biblical text in many editions sponsored by Catholics. However the apocryphal books were removed by Protestants from the Bible published under their aegis. In spite of the absence of religious sanction for the holiness of Second Esdras, it remains a powerful work which informs us of at least one mode of prophetic behavior which might lead the prophet to an ecstatic state in which visions and voices could be heard.

Ezra tells us he is a direct descendent of the tribe of Levi, the tribe assigned the duties of helping the priests in the Temple. Ezra, like Moses and Jeremiah, feels unworthy to be the Lord's messenger. Again the Lord insists. Ezra castigates the people for sinning and describes death, destruction and exile as their just punishment. But Ezra is a logical man. He asks the Lord why he, Ezra, cannot be the vehicle through which the Lord's words transform the people. They would then be spared from the punishments spelled out by God. Through a series of meetings with an angel of the Lord, God makes clear to Ezra that Ezra's role is not a redemptive one but merely a prophetic one. Ezra is simply ordered to tell it to the people as it was then and is to be in the future. Ezra challenges the Lord: what is the sense of creation if the Lord is to destroy so many of those He has created? Then, Ezra raises what have become, and clearly already were, age old questions: why is there evil in the land and why do the bad triumph while the good suffer? How could God permit this to happen? As He answered Job, so God answered Ezra: the mind of God is not for man to understand. Reacting like any man, Ezra is unhappy, even desperate, when he hears such things. He wants to know who will ultimately be saved and given a life in heaven and who will not. He is shocked to hear that God made Earth for the many but that Heaven is only for the few. Even more shocking is news that is not consistent with other passages concerning salvation; passages that scholars believe may have been late interpolations in what became a multiauthored book. The shocking news is that repentance will not guarantee a trip to that small Heaven. Who shall go and who shall not has been preordained and cannot be known in advance. A similar doctrine of predestination was considered by Luther and Calvin, founders of Protestantism almost 1500 years later. These ideas did not, ultimately, become part of the official thinking of most Christian denominations. As Ezra logically complains, why be good if one cannot be guaranteed a place in Heaven by that behavior?

The author of Ezra is thought to have been Jewish. In his text, predestination on one hand, and, on the other hand, heaven as the motivation for good behavior either reflect very late developments in Jewish thought or later Christian additions to the text. As I have emphasized in earlier sections of this book, an important aspect of Judaic ethics was not the reward of a salvation in heaven but rather the call to do good for its own sake and for the sake of the society that benefits from each individual's good behavior. Ultimately, for the Jews, the reward for good behavior was to be God's continuing support for their descendants and fulfillment of His covenant.

Notwithstanding discussions of salvation, God, when speaking through Ezra, once again instructs us to champion the widow, defend the orphan, give to the poor, cloth the naked, care for the disabled and shelter both the very young and the very old. But I speak of Ezra in order to make another point. Ezra serves as an example of a prophet who may have used artificial means to bring about a hallucinatory state of ecstasy in which he then heard voices, had visions, and through these became motivated to fight the uphill battle for the good of humankind.

The text in the Apocrypha tells us that Ezra is required to fast repeatedly, each time for 7 days in order to bring forth the angel that speaks to him. Those of us who have become light headed after fasting for only one day can readily imagine our mental state after one week. Surely one may no longer be "in their right mind".

But Ezra does not only fast. On one occasion he is required to eat in order to hear his angel. But what is he required to eat? He is told to go to a field and to eat only what grows there and to pray unceasingly. Then he will be visited. What grew in that field? We are told that flowers grew in the field and that Ezra ate them for a week to his heart's content. Then he eats in the same field for another week after which he hears voices, has visions and feels himself transported to the Nations to present them with his final apocalyptic message.

What kind of flowers grew in that field? Poppies perhaps? Or some other hallucinogens? We will never know. But we may certainly suggest that they were, indeed, hallucinogens. Ezra's experience only occurred after two weeks of eating the flowers of the field. This passage provides an important clue supporting the suggestion that the prophets may have used mind altering drugs to bring forth their visions and voices.

If fasting or hallucinogens do not explain prophetic visions, what does? One initially less palatable choice remains—mental illness.

Is Mental Illness A Possible Explanation For The Experiences Of Some Of The Prophets?

The answer to this question depends, first of all, on our definition of mental illness. For example obsessions are a form of mental malfunctioning. Some persons with obsessions seem well aware of the practical problems caused by their obsession and may struggle to avoid the obsessive behavior. The stronger the obsession the less likely the success of the struggle. Jeremiah tells us that he strives mightily to avoid going out among the people and among their leaders to preach his message and to make his prophecies. He describes this struggle in terms of his arguing and pleading with God to relieve him of the burden. But God persists and Jeremiah does that which he dreads to do. Could it be that the description of this struggle is simply the description of an obsessed individual struggling with his personal demons in an effort to avoid placing himself in the politically dangerous position that threatened his life?

Religious obsession and fanaticism are psychological cousins. The second year of the twenty first century shows how greatly a religious fanatic can influence the thoughts and behavior of masses of people. In this case, the fanatic is Osama Bin Laden. Perhaps, in his case, some of his influence is a product of the schooling which exposed so many Islamic children to anti Western sentiments and to the ideas of violent Jihad and martyrdom as the ultimate in righteous behavior. For the West the question is whether Bin Laden is "crazy" or merely evil? Perhaps not enough is known about Bin Laden to answer the question. In part the answer depends upon whether its subject claims to have sensory experiences that others cannot share. Such experiences are called hallucinations. They are experiences that the individual believes are real even when others are excluded from the experience. Such experiences fail the reality testing applied by the community. As such they constitute symptoms recognized by the medical community as indicating one of the psychotic mental illnesses. Certainly the prophets reported such experiences.

But when we consider whether some of the prophets were psychotic we are forced to look at a series of questions. Can a psychotic function well enough to organize the activities of daily life? Can a psychotic think logically about things outside the particular delusional system that has captured him? Can such a person maintain a highly developed ethical sense? Can that person communicate to an audience in a manner coherent enough to impress and influence them? Would the mentally ill function well enough to command the respect of the community; respect resulting in the preservation of their words for future generations to hear and to read? Would the mentally ill function well enough to be politically influential. Could they become so influential that they threatened the establishment; a threat so great that in the cases of Elijah and of Jeremiah their respective kings threatened them with prison and death?

The answer to all these questions is yes. A very well documented example is that of Daniel Paul Schreber.

Evidence From The Schreber Case

Schreber was born in Leipzig, Germany in 1842. His first episode of recognizable mental illness occurred while he was presiding Judge in a lower German court. Upon his release from the mental hospital he took up a post in a higher court. Nine years after his initial hospitalization and only one month prior to his second hospital admission he took up a position as presiding judge in the court of appeals of Saxony. This was the appeals court in Germany's second largest state.

As we shall see, Schreber's illness was not intermittent. He earned respect from and advanced within the legal community while he was ill. During the nine years between hospitalizations, his delusional system became more systematized. Moreover, during the years that followed, he maintained his ability to reason, to write logically, and to display a highly developed sense of ethics. These traits are manifest in the book he published in 1903. That book, Memoirs of My Nervous Illness, made his case famous and was the subject of Freud's essay titled Psychoanalytic Notes on an Autobiographical Account of a Case of Paranoia (Dementia Paranoides), published in 1911.

Schreber's delusional system involved God and Schreber's relationship to God. Schreber tells us of "the certainty of my knowledge of God and the absolute conviction that I am dealing with God and divine miracles." He denies the applicability of any scientific test of his beliefs, saying that his certainty in the beliefs

". . . towers high above all human science". He believes that ". . . insight into the true state of divine matters was granted me in an incomparably higher degree than any other human being before." He acknowledges that for this "insight" he has had to "pay dearly . . . with the loss of my whole happiness [years of incarceration in a mental institution] for a great many years". Schreber writes these statements in 1901 in an appeal to a German court, petitioning for his release from the mental hospital. In this appeal he argues that he is not a danger to himself or others. He says that he can take care of himself as well as any other man and that what others call delusions have had no negative impact upon these functions of everyday life. He also points out that no one questioned his legal decisions during the years prior to incarceration. His argument is persuasive enough to achieve the end to which it was directed; he is released. He does not return to the court, but so far as is known he is able to maintain himself as he predicted until 1907. Then his mother dies and his wife has a stroke. He is admitted to a mental hospital shortly thereafter and dies there four years later.

As Freud succinctly describes it, the court judgment that gave Schreber back his liberty, recognized that he still believed that he had a mission to redeem the world and restore it to its lost state of bliss. But a world catastrophe must first occur which would necessitate Schreber's recreation of the species through the gift of God. Schreber tells us that he could do this if he were both male and female. Schreber describes periods in which he received poorly formed female organs. He says that these two periods were analogous to the conception of Jesus in a Virgin body. Schreber acknowledges that he is ill. But, he says, the illness is nervous and not "mental". He says that he really is beset by hypersensitivity of his nervous system. This gives rise to the symptom that particularly dictates his removal from society; an apparently uncontrollable vocalization; the sudden and frequent wailing of loud, unpleasant, and wordless sounds. But he insists that God is really assaulting his nervous system. His symptoms are not the sign of mental illness but are the normal response of a nervous system assaulted by real experiences that are simply not shared by others. These assaults upon his body are not delusions. These real assaults are the cause of his nervous illness. These causes he calls "miracles". They include a worm in his lungs, compression of his chest, the pouring of food and drink directly into his abdominal cavity and thence into his thighs, the sawing and sometimes the perforation of his skull with the pulling out of his nerves. Schreber also tells us that the sun—a female figure in his delusional system—has spoken to him for years with a human voice thereby revealing herself to be either a living being or the organ of a higher being. The sun and rays from the sun figure prominently in Schreber's delusional system. He sees these rays as responsible for the visions of Joan of Arc and for other historical miracles. He distinguishes between pure rays and impure rays. They penetrate him with diverse effects. They may also be transformed into visual images. He is quite aware that all of this is supernatural. But that is exactly his point-this is not insanity but rather the product of miraculous intervention. These experiences are all produced by God in accordance with a plan. It is not necessary to continue. The reader can clearly see the relevance of the Schreber case to the question asked here. The successful existence of the psychotic in the general community depends upon the type of psychosis. Clearly some psychotics like Schreber can function for years in highly respected and influential positions. Clearly they can believe that what we call their delusional system can put them in an intimate relationship with God and with God's plan for humanity. But why the particular content of the messages that they receive or the thoughts that come to them while they were delusional?

Genes Can Explain The Prophets,Their Message and Their Receptive Audience

Why the ethical dictums that became the heart of Judaism? Thus far I have suggested that the message of the prophets came to them in dreams or as the product of hallucinations and delusions that were induced either by drugs, environmental deprivation or mental illness. I made these suggestions because of the fantastic nature of some of the prophetic texts and because of the collateral evidence that I presented in order to show that such alternatives are possible. However, there is, of course, a more straight forward explanation. The message may have come to its originators and propagandists without the presence of mental illness and without the use of drugs, fasting and desert heat to produce a state of ecstasy. It may simply have been a product of ethical genius. If the ethical precepts were imported from Egypt they had to be internalized by the prophets and repeated again and again to the Jewish people. The "announcers" had to be expressing convictions deeply held and deeply internalized. Persons who take on this role, unwanted and dangerous as the Scriptures suggest, must certainly be a very special kind of human. It is to them that I apply the term "ethical genius".

Their thoughts and words must be dependent upon the chemistry of their brains. In the next chapter I will explain that all mental behavior is a product of the brain's chemistry and that this in turn is totally dependent upon our genes. I will explain how environmental influences can only produce the mental experiences which the genes permit. I will explain how environment, working through the genetically encoded chemistry of the brain, can alter that chemistry to promote new images and thoughts, but that environment only does so in accordance with the genetically determined machinery that is already present. Genes can explain not only the thoughts and actions of mentally healthy individuals but also explain the thoughts and behavior of those affected by hallucinations or delusions.

CHAPTER FIVE

GENES AND ENVIRONMENT

I propose that genes are the ultimate explanation for the World view transplanted perhaps from Egypt into the Middle East. If transplanted then this code was brought to a final form which was accepted under the aegis of a monotheistic God as the ethical code of an identifiable people and remains so to this day. Moreover that code was adopted as the ethical underpinning of each of the successive monotheistic religions that followed Judaism. In order to understand the role of genes in history it is necessary to understand what genes are, how they affect what we call the mind, and what the relationship is between genes and environment.

What Are Genes?

Every organ of the body is made up of structures called cells. Every cell contains structures called chromosomes. The chromosomes contain strings of chemicals called DNA. We need not concern ourselves with the longer name represented by that abbreviation. Regions within the DNA are genes.

What Do Genes Do?

Genes provide a code which instructs each cell of the body to make certain proteins. Some proteins are made by virtually all cells, no matter where we find them. But some proteins are made exclusively by one set of cells rather than others. That is because some genes are "turned on" and others are not. The reasons for this are not essential to this discussion. When a gene is not turned on we may say that it is not "expressed". The different expression of genes in different places within the developing embryo accounts for the differences between an arm, a leg, a liver, a heart, a brain, and so on. Under the direction of the genes, the cells of the brain assume their shape and function. All of this is the result of diverse proteins made according to genetic instructions.

Genes and Brain Function

The brain contains several kinds of cells. The nerve cells are called neurons. Some genes in neurons are turned on to produce the proteins that give neurons their shape. Other genes in neurons are turned on to direct the production of the proteins which in turn control the manufacture of special chemicals called neurotransmitters. Nerve cells in different parts of the brain are programmed to produce different transmitters.

Neurotransmitters are released from the neurons that make them. These chemicals then travel across the very small space between themselves and their neighbors. Thus the neurotransmitters reach other neurons. The neurotransmitters may then trigger these receiving nerve cells. By this is meant the generation of an electrical current in the receiving cell. The ability of neurotransmitters to trigger other neurons depends upon the presence of specific recognition sites on the receiving neuron. These recognition sites are called receptors. The receptors are also made under the direction of genes.

When neurons are triggered, this is accompanied by the altered manufacture of proteins or the altered configuration of proteins. The electrical current generated in the receiving nerve cell may lead it to release neurotransmitters of its own. Thus a chain of neural signals may develop, all in less time than it takes to read a sentence or even a single word.

Some of the physical or chemical changes remain as "traces" of prior experience and account for learning and for memory. Some of the physicochemical changes result in what we call conscious experience. Such experience includes things remembered, dreams, thoughts, emotions and all sensory experiences such as seeing, hearing and feeling. In short, the conscious and unconscious experience that has been called the "mind" is simply a manifestation of genetically controlled chemical and physical events within the brain. The phenomenon of the mind is, just like the phenomenon of sight or hearing, an experience reflecting underlying chemistry that is controlled by genes. Evidence in support of this belief comes from two sources: studies of humans and more recently begun studies of animals. Let us look first at the animal data.

Control Of Behavior By Genes— Evidence From Animal Studies

There are now animal studies that show the relationship of genes to behavior or "personality". The animal studies can be carried out in ways that are not possible in humans. One can manipulate the egg of the female in ways that result in the insertion or deletion of a preselected gene. Mice have been the animal of choice for such studies and the resulting "transgenic" mice are called "knock-in" or "knockout" animals depending on whether a gene has been added or removed. Behavioral profiles of such mouse mutants are being compiled. Genetically modified mice have been developed which demonstrate aberrant social, reproductive and parental behaviors or learning and memory deficits, or aggression, or alterations in behaviors thought to reflect the equivalent of anxiety in humans. In other words it is beginning to be possible to investigate the relationship between specific genes and highly complex behaviors. Some of these behavioral traits are analogous to what we call "personality" in humans.

New biochemical tools are available to facilitate this kind of study. Among these techniques is one that enables the rapid, simultaneous identification of large numbers of genes. This technique provides insight into the complexity of gene interaction and its relation to behavior. That is to say that many genes, rather than a single gene, may be necessary for a behavior to be displayed or for there to be a propensity for displaying a particular behavior.

As a beginning, such studies examine animal behaviors that are easy to observe and measure. One example is running or exercise. Many readers have seen gerbils using an activity wheel placed in the animal's cage. Mice will also use such a wheel. It has been discovered that some mice use the wheel much less than others. Over 25 different genes have been found to be disproportionately expressed in the low-level runners. These genes appear in a far greater number of these mice than in the high level runners. On the other hand, more than 15 other genes are found to be preferentially expressed in the high level runners. Like all genes, each of the genes in question has a relationship to biochemical or biophysical function. These functions are known for many of the genes in question, but in several cases the function of the gene has not been identified. The way in which the identified biochemical and biophysical functions relate to the complex behavior of the "runner" or "non-runner" is not known. Whether such a complex system can ever be understood remains to be seen. But we should not confuse lack of understanding with the absence of a relationship between the discovered genetic differences and the behavior which they parallel. The relationship clearly exists, whether we can explain it or not. Further evidence for a relationship between genes and mental behavior or personality comes from studies of humans.

Evidence From Studies of Mental Illness

Studies of the mentally ill provide another source of evidence showing the control of mental behavior by genes. These studies show that genes can constitute up to 30 or 50% of the determinants of some mental illnesses. Since the influence of heredity is a far from 100% the reader may immediately think that there are non-genetic determinants of mental illness. These other determinants would be environmental. But, as we shall explain, even the influence of the environment is genetically determined.

The most reliable of the genetic studies of mental illness use twins as their subject. Twins have been studied in the home of their birth parents and in adoptive homes. If one twin has a mental illness the chances are increased that the other twin will develop the illness. The occurrence of disease in both twins is called concordance. Identical twins have a much more identical genetic makeup than fraternal twins. The genetic influence on the illness is shown by the fact that the concordance is greatly increased if the twins are identical rather than fraternal. Identical twins have much more identical genes than do fraternal twins. The relationship of concordance to the genetic similarity of the twins is found not only when the twins are reared in the same home but when they have been separated at an early age and reared in separate homes. Thus the genes, not the home environment, were the most important determinants of the development of the disease.

The development of mental illnesses has also been examined in the light of genetic relationships other than twinship. Parents with a mental illness are more likely than other parents to have children with that illness. When there are two parents with the illness, the children are much more likely to develop that illness than when only one parent had the illness. In general, persons with mentally ill first degree relatives are more likely to share that illness than are persons with mentally ill second degree relatives. These facts hold true irrespective of the homes in which the involved persons have been raised. These facts are valid not only for illness but also for personality traits in general. Personality traits between twins develop along very similar paths even when the twins are reared apart. The similarity of the traits is much higher in identical than in fraternal twins. To quote one study by Tellegen and coworkers: "Consistent with previous reports, but contrary to widely held beliefs, the overall contribution of a common family-environment component [to personality traits] was small and negligible for all but 2 of the 14 personality measures."

Data reviewed by McGuffin and Thapar confirm this. Among the traits so strongly shared by close relatives, in spite of the absence of shared environment, were self reported extrovertism and self reported neuroticism. A similar phenomenon holds true for other aspects of temperament, for occupational and leisure time interests and for social attitudes.

Moreover, gene products interact in complex ways. Indeed, one must look not only for a genetic condition causing disease but also for the absence of genes which help ward off the disease. In this vein, geneticists and psychiatrists have begun to investigate the existence of "wellness" genes whose differential distribution might account for the different responses of different people to identical environments

Some genes are "dominant" and others may be "recessive". When a gene is dominant, the contribution of the gene by a single parent may be sufficient to observe the genes effect. In the case of recessive genes, each parent must have the gene and each must transmit it to the child in order for an effect to be manifest. But sometimes it is not a question of transmission of the whole gene that is at issue. Rather there may be very tiny changes in a gene which are responsible for an altered function. Such changes are called polymorphisms. These minute alterations within a gene can alter behavior. For example one such alteration has been associated with aggressive behavior in people with Alzheimer's disease. We will revisit the importance of gene polymorphism later in this chapter when we examine the interaction of environment with genes. We will also talk about polymorphism in the next chapter which concerns the existence of "Jewish genes."

But if everything we are and become is under genetic control then why is it that twins, even identical twins, brought up together may still differ in the way they behave? And why doesn't mental illness, when it strikes one twin, always strike the other twin? The first answer to this question is a simple one: the environment may not be identical for both twins even when raised in the same house. Even identical twins growing up together do not have identical environments. Differences begin at birth since one twin is born before the other. But we must keep in mind that the environmental stimuli can only act as genes permit them to act. The genetic dependence of environmental effects will be explained shortly. For now, simply keep this permissive effect of the genes in mind. Then, think of the small, chance differences between the ways in which each twin is handled or between the ways in which other people respond to each twin. These small differences will have effects on subsequent mental functioning. Later in this chapter I will explain how small initial effects on brain function and development can be explosively magnified as new neural circuits in turn effect development of subsequent circuits. Thus, small initial differences in environmental stimuli can ultimately have significant effects on subsequent responses to new environmental stimuli. However this answer to our question does assume that one twin was born with the same genetic material as the other twin. This assumption may not be correct, even for identical twins. To explain this, we must turn to relatively new discoveries about genetics.

Genes Are Not Distributed Equally to All Cells Of The Body Even In Identical Twins

Thus far, the genes we have considered are all located within the center of each cell, on the chromosomes within a structure called the nucleus. The bits of genetic material on chromosomes are evenly divided between all the body's cells. This division is essential to our understanding of the genetics we may have been taught in school, especially prior to the end of the twentieth century. However, we now know that not all genes are found on the chromosomes, Small amounts of genetic material (DNA) are carried by cellular components called the mitochondria.

Some people believe that mitochondria or mitochondrial genes were introduced into cells many millions of years ago when life forms were very primitive. This theory suggests that simple forms of animal life may have been parasitized by other single celled organisms which took up residence outside the nucleus of the occupied cells. The occupier and the host cell existed in a happy, symbiotic relationship and eventually the DNA of the occupier became an essential part of the host's machinery, incorporated into the mitochondria.

This theory may or may not be correct. What is correct, is the fact that mitochondrial DNA [mitochondrial genes] are present in the cell and may provide a partial explanation for the fact that environmental influences do not always appear to be dependent upon the nuclear DNA transmitted by egg and sperm. Nuclear DNA is carried to all cells of the body as the cells of the embryo divide and the fetus is formed. But mitochondrial genes are not equally divided between all the body's cells; they are apparently randomly distributed as the embryo develops. Thus some mitochondrial genes may wind up in brain cells in one person and not in another. This could account for behavioral differences even in identical twins brought up together. Moreover, not all brain cells may have the same complement of mitochondrial genes. This could also account for different functions in different brains. Mutations on mitochondrial genes might also account for variations in behavior.

The proportion of brain cells containing aberrant mitochondrial genes may differ from person to person, even from one identical twin to another. The effect of a mutation in these genes depends upon the cells to which the mutant DNA is distributed. It also depends on the number of affected mitochondria within those cells. Thus the effect on brain function would depend on how many altered mitochondria are present in a neuron and which neurons are affected. Since distribution of such genes to various cells in the body is apparently random it is entirely possible that even in identical twins there are some genetic dissimilarities that enable neurons to choose divergent pathways for action. This, and not environmental differences, could account for the absence of 100% concordance of a mental illness or personality trait in identical twins. At present such an explanation is only hypothetical. However psychomotor regression and psychomotor retardation have already been associated with some mutations in mitochondrial DNA.

Modern genetics also provides another hypothetical explanation for differences in mental behavior of twins; differences that might otherwise be ascribed to some variation in the way they were raised or to other aspects of their environment. It is now known that some genes within the cells nuclei develop an instability which leads portions of the DNA to replicate in a long chain of so-called "triplet repeats". These chains are replications of the same three part piece of genetic information. Some neurological and behavioral diseases are now known to be caused by such repeats. The chains lengthen during life. Disease occurs if the chains reach a certain length in cells critical for the function of the nervous system. Repeats occur on different genes or chromosomes in different diseases. One disease caused by repeats and characterized in part by disordered mental function is known as the fragile X syndrome. This is the most common type of mental "retardation". It is now known that the tendency for repeats to form is not distributed equally to identical twins. The reason for this is not understood. In fragile X syndrome one identical twin may be affected by repeats and manifest severe symptoms while the other has neither the same repeats nor the same symptoms. A similar phenomenon rather than differences in environment could account for the expression of other forms of mental function in one twin but not in the other. Attempts to find important triplet repeats in the mentally ill or associated with differences in behavior have just begun and therefore we do not know the extent to which different development of triplet repeats can account for different development of mental illness or development of personality differences in otherwise genetically identical individuals

I hope this brief discussion of new discoveries about genes has provided additional support for the claim that our genetic material can determine all that we are and that genes mediate the connection between our environment and the development of our mind with all its senses. It may further our understanding of this assertion if we look at our interaction with some important and obvious external stimuli that are often omitted from our concept of the environment.

Environment Acts Only As Genes Permit

There are three ways in which the environment can either modify genetically determined responses or effect the way genes express themselves. The first of these is quite conventional. Environmental inputs can mutate the gene. To understand this on the simplest level let us first look at an example which has nothing to do with the "mind". Let us look instead at environmentally produced cancers.

Two of the environmentally related cancers are well known to all of us. They are lung cancer and skin cancer. Tobacco smoke is an important cause of the former and sunlight of the latter. How does this happen? It happens because these two environmental insults injure the DNA of lung and skin cells. The genes are portions of the DNA. The cells of cancer victims have been unable to repair the injuries to these genes. The injuries alter genes in ways that favor the unrestrained multiplication of cells and the invasion of adjacent tissue by these cells. This unrestrained growth and the invasive properties of the multiplying cells is what we call cancer.

Now let us return to the cells that are the substrate of what we call the "mind". These cells are the neurons. The question we ask is whether there are known environmental inputs that can mutate genes that control mental activities. At present there are no definite examples of such interactions other than the actions of poisons that can destroy neurons. Examples include lead and mercury. Exposure to toxic levels of these metals certainly can alter the anatomy of the brain and hence effect behavior. Recently there has been much attention played to the role of vaccines in the production of autism. However the best science on this subject has so far been unable to prove a link between the two.

But there are other ways in which ingested substances—environmental inputs—can change genetically determined mental behavior. Drugs of many kinds can do this. The genome is not altered but the drug impacts a neurochemical pathway that leads to a change in the genetically determined event. The drug only works because there is a preexisting genetically determined neurochemical path that the drug is able to target.

The preceding examples—neurotoxins and drugs—are rather trivial examples of environmental influences on behavior. The drug—for example a serotonin reuptake inhibitor in the Prozac family of drugs—may change the abnormal balance of a neurotransmitter in a genetically influenced mental behavior like depression or obsessive compulsive disorder, thereby ameliorating the symptoms. But the drug does not alter the gene[s] or the gene translation responsible for the disorder in the first place. However we now know of ways in which environment can actually alter gene expression and do so without mutating the gene. This potential path to behavioral modification via the genome was unknown until the late 20th century and depends upon what is called epigenetics.

Genes themselves are classically defined as stretches of DNA that code for protein production. The gene first leads to the production of messenger RNA that forms a template for the synthesis of a protein specific to that particular stretch or DNA [i.e. a gene]. However we now know that the vast regions of DNA between genes are not "junk" but contain regions that code for small molecules of RNA that can control whether the nearby genes are turned on or off. The production of these small RNAs can be altered by changes (polymorphisms) at single points in the non-coding portions (previously called "junk") on the DNA. The production of these small RNAs can also be altered by altering the chemical groups attached to the DNA previously thought of as "junk". Scientists now know that environmental factors can lead to the chemical alterations of the DNA that codes for the small RNAs. It is even possible that the body's chemical response to such things as stress leads to the chemical modifications of the gene regulating stretches of DNA. Moreover the DNA-gene and non-gene—is wrapped in a shell of protein the configuration of which determines whether a particular gene can operate. Chemical alterations of this encasing protein can alter the way in which the genetic DNA is made available for the production of messenger RNA and hence for the ultimate synthesis of the protein coded for by the gene. Again these chemical alterations can be altered by environmental factors. Epigenetics is the name given to the effects on gene expression produced by non-genetic DNA or by effects of the configuration and chemical modification of the protein encasing the DNA. Epigenetic events provide a path by which the environment can directly modify gene expression without causing gene mutation. and without merely effecting a neurochemical pathway that can change the behavior that is dependent upon some competing, neurochemical event.

It is important to note that different individuals may respond differently to identical environmental events. Drugs and talk therapy are environmental inputs. But these therapies affect different people in different ways. This is so because gene expression differs among brains. Thus neuronal chemistry in one brain may be more resistant to "talk" therapy or to a given medication. These differences between brains begin at birth, much as skin or eye color differs among people at birth. Because such differences can affect the way in which we respond to drugs, there is now a new branch of science called pharmacogenetics which investigates how genes alter our response to drugs.

Genetic differences present at birth are only the beginning of the story. Unlike fixed traits like skin color, the brain has a number functions that are malleable or "plastic". The plasticity of these functions accounts in part for recoveries after brain injury. Plasticity appears greater in the young than in the old, but is present in everyone. This plasticity is an expression of the brains ability to make new proteins or to form new receptors and hence new or altered signal pathways between cells. Since the initial capacity to do this is genetically determined there will be some difference in this capacity among individuals at birth. Then the environment, responded to in a genetically determined way, produces additional physico-chemical changes within the brain. These changes then alter the way in which additional environmental stimuli are responded to. As new transmitter- receptor pathways develop, behaviors, thoughts and emotions will develop and change.

In some instances there may be a snowball effect, greatly exaggerating differences that were once small. Thus, as our lives go on, the responses of the brain—the mental phenomena or so-called "mind" that reflect brain chemistry—may diverge even further among a group of people. This will be so when people with different brain gene expression are exposed to identical environments. Divergences will be even greater when the stimuli impinging upon genetically based brain activity are stimuli coming from different environments Responses to drugs can provide good illustrations of the snowball effect that gene-directed pathways may have on responses to an environmental stimulus. Addicting drugs trigger release of a chemical in the brain. Some neurons have receptors for that chemical. The more a drug is used, the more the neurotransmitter is released. The receptors for that transmitter appear to adapt by increasing their number or by increasing their avidity for the transmitter. This is a form of the "plasticity" mentioned above. This alteration is out of proportion to, or causes other chemical events out of proportion to, the increment in drug intake. When the receptors for the transmitter released by the potentially addicting drug go unfilled this leads to the craving sensed by the addict. Craving results from the triggering of complex pathways that "recognize" unfulfilled receptors; this craving is relieved when the receptors are "satisfied". Satisfaction occurs when the receptors are fully occupied by the chemicals whose production is triggered by the drug to which the addict is addicted. Genetic differences among individuals account for one or more differences along the pathway that begins with the number of receptors, and continues through transmitter release and receptor number and adaptation. Thus, a small genetically controlled difference along this pathway may make one person susceptible, and another not, to the development of a drug craving which will then be intensified by the snowball effect just described.

It is now possible in living persons to actually see the binding of certain drugs to their receptors. This is done with an imaging technique analogous to making X-ray pictures but obtained with a different technology. From the location and intensity of binding we can see that addicts have more of a certain type of receptor than non-addicts or, at least, that the receptors of addicts are "turned on" to a greater extent. This effect is long lasting and may account for the difficulty in ending addiction. Scientists are attempting to develop drugs which will fill the receptors and dampen the craving without setting up a new craving for the therapy itself. The alternative is a "talk" therapy or attempts at behavioral modification which might, through complex interactions with other neurochemical pathways, alter the number of unfilled receptors or dampen their neurochemical effect. Addiction is a reflection of what we call the "mind". As scientists learn more we will be able to propose ways in which all sorts of differences between the "mind" of one person and the "mind" of another may be explained and modified through drug treatment and we will gain greater understanding of how the effects of treatment are dependent upon an individual's underlying genome with its polymorphisms.

A Word About Genetic Determinism

If we are what our genes permit us to be, then would it be possible for scientists to someday predict exactly what an individual will do and predict accurately at all times? Such predictions are the ultimate outcome of what has been called genetic determinism—the idea that our behavior or mental state is predetermined and cannot be altered. As I have indicated there is reliable data showing that personality traits and mental diseases can, in fact, be predicted with greater than chance accuracy if we know the pertinent genetic information about the individuals in question. But, as I also indicated, genetically determined pathways are frequently and perhaps continuously altered by environmental inputs and there is therefore a large measure of error when we predict how an individual will respond or think in a given set of circumstances. Presumably our accuracy would be greatly increased if we could perform a sort of biochemical dissection of all the relevant pathways in the individuals brain because this would provide us with a map that reflects both the original genetically determined paths and the effects of environmental inputs on those paths. It is doubtful that such knowledge will be available in the foreseeable future. And, in any case, there will always be the possibility that an individuals future behavior under the same circumstances may be altered by later environmental inputs, working as permitted by that individuals underlying genetic capacity.

In addition to modification by the environment, there is another way in which genetically determined mind-work can be modified, and in this case become unpredictable. It is possible that random neurochemical events affect final outcomes. It may be that certain neural pathways frequently fire, randomly and with no necessary result. They are not "heard" by receptive circuits elsewhere in the brain. It may also be that small amounts of neurotransmitters are released randomly at various sites in amounts that do not normally trigger their receptors. Now let us imagine the sudden intrusion of some other brain activity, triggered by an environmental stimulus. The effect of that trigger may depend on which randomly firing circuit happened to be firing or releasing transmitter at that time. The combination may result in effective excitation of pathways leading to anger or might instead stimulate pathways leading to reconciliation. All the pathways are genetically determined but because some were firing randomly it is not possible to predict whether the final result will be peace or war—it would depend on which randomly firing circuit happened to be active at the time of the triggering environmental event. Therefore we may not have absolute predictability even though all pathways are the result of genetic control of protein synthesis responsible for the structure and function of the brain

Thus far, I have tried to explain the relationship of mental function to the product of our genes. I have explained how brains that are different at birth may become increasingly different as time goes on. I have explained how genes control different responses to the same environment and different responses to different environments. The personality that reflects underlying changes in chemistry, is also always a reflection of the chemical events permitted by that individual's genes. Having come this far we must now discuss the potentially vexing question: is there such a thing as "Jewish genes"?

CHAPTER SIX

JEWISH GENES

Definition

By a Jewish gene, is meant a gene that is carried in a disproportionate degree by persons who believe that their ancestors were Jewish. Such genes will also be present in a disproportionate degree in apostates from Judaism and in their descendants.

Evidence From Incidence of Diseases

Defined in this manner, the existence of Jewish genes has been known for some time. For example, a look at any comprehensive medical text book will reveal a number of rather rare diseases which, when present, generally manifest themselves in Jews who trace their ancestry to the group that moved to central Europe from ancient Palestine. These Jews are termed Ashkenazi Jews, the word Ashkenazi meaning "from a Biblical people" and originally applied in the ninth century to Jews of German origin.

In the brain, the most common of these diseases was Tay Sachs disease. This devastating inherited disease is manifest in infancy and proceeds through stages of increased deterioration of brain function and on to death.

Genes come in pairs, one on each of a particular pair of chromosomes. One chromosome in a pair is contributed by the father and one by the mother. Tay Sachs disease requires that both chromosomes in the relevant pair carry the defective gene. Therefore, to conceive an affected child, both mother and father must carry one copy of the defective gene, and transmit it, rather than its healthy pairmate, to the fetus. People with a single copy of the defective gene are called carriers of the disease. Genetic testing of the potential parents permits identification of carriers. This in turn alerts parents to the possibility that they may conceive a child with the disease. When the parents receive such a warning they may elect not to conceive a child, or may elect to have tests of amniotic fluid after conception in order to avail themselves of an abortion should the fetus have the disease. Even when prenatal genetic testing has not been performed, tests of amniotic fluid will still permit identification of an affected fetus and permit the parents to elect an abortion to prevent the birth of an affected child if that is their wish. Prenatal testing and therapeutic abortion after genetic testing of amniotic fluid have resulted in virtual disappearance of the disease. Many other heritable diseases can be abolished in this way and the gene pool of Ashkenazi Jews will certainly be changed by removal of the diverse diseases that have existed with disproportionately high frequency in this group of people.

Other diseases have distributed themselves in disproportionate degree in Sephardic Jews. In contrast to the Ashkenazi subgroup of Jews, the Sephardic subgroup traces its ancestry to Jews who fled from ancient Palestine to the lands abutting on the Mediterranean and particularly to Spain after being driven from, or voluntarily leaving their original ancestral home in ancient Palestine. The rare diseases shared particularly by this group of Jews include familial Mediterranean fever and Machado Joseph disease. The latter is named after a Portuguese sailor who brought the disease to America in the nineteenth century.

Obviously, the preponderance of certain genes in certain groups of people is a result of intermarriage within that group. Traditional prohibitions against marrying out of the religion, plus the effects of prejudice against Jews among the peoples in which they lived, are causes of this intermarriage.

But marriage between members of the same group not only accounts for the general distribution of certain genes throughout either Ashkenazi or Sephardic Jews. Such marriages also account for the distribution of certain Jewish genes within other subgroups of Jews. Examples of such distributions concern altered genes on the Y chromosome.

The Y Chromosome And Jews Of The Priestly Line—The Cohens or Kohanim

For the most part the genes are stable over many millennia. But occasionally, for a variety of reasons, slight changes occur. Sometimes these alter the function of the gene, sometimes they do not. Alterations in the DNA in and around genes are called polymorphisms. Polymorphisms on the Y chromosome will only distribute themselves among Jewish men since the Y chromosome is present only in men and determines that the embryo becomes a boy. Now let us imagine that a particular change [polymorphism] or set of changes took place on the Y chromosome at some remote time in antiquity. Let us imagine too that by chance the first person to accidentally develop such a change was identified by a particular name. Finally, let us imagine that that name in one form or another is carried down through subsequent generations of men—the children, grandchildren, great grandchildren, etc. of the originally affected male. In such a case, these subsequent generations of men should all show the altered gene while those not descended from the original father figure should not have the altered gene. We can also imagine the inverse of the preceding scenario. If polymorphisms occur in persons who are excluded from a line of descent, then only descendents outside of the excluded line will have those particular genetic signatures. Something like this appears to have actually happened to the Jews as revealed by a study of those who bear the name Cohen. Let us look at their story.

Jewish men who consider themselves members of the Cohen "clan" are often but not necessarily named Cohen. They believe that they are descended from the priestly line that, according to the Bible, began with Aaron, the brother of Moses. If there were a semblance of truth in this story then present day Cohens might have genetic signatures carried down from the first priests through successive transmission of the Y chromosome. These signatures may have had no causal relationship to the priestly proclivities of the Aaron figure. They were probably merely tags, developed by chance, which help identify his pedigree. The more matings that occurred throughout the generations the more widely the genetic change would be distributed. All other Jewish males should have a slightly different DNA, stemming from the slight differences in the DNA of the Y chromosome that already distinguished the first priest from the non priests. Moreover, if other genetic changes occurred among Jewish males outside of the priestly line, then these changes would tend to propagate with the Y chromosome donated by these nonpriests to their male progeny throughout the generations. In fact, analysis of Y chromosome patterns has shown two "signatures" with significantly different distributions in Cohens vs non Cohens. That signature was found in 18% of non Cohens but was present in only 2% of Cohens. Moreover the "signature" within the Y chromosome of Ashkenazi Cohens—the absence of the more frequent DNA pattern—was the same as that in Sephardic Cohens. This indicates that the original founding member of the clan preceded the split between the Ashkenazi and Sephardic populations. In other words, the genetic change occurred before the massive dispersion of the Jews that occurred in Roman times following the last of the devastating wars between Judea and the Roman Empire. This conclusion is bolstered by additional studies of Y chromosome alterations in Cohens and non-Cohens. Indeed calculations based on known rates of modification in DNA and an analysis of the degree of difference between the DNA of the Cohens and non Cohens led to the conclusion that these two gene pools could, indeed, have resulted from a separation of ancestry approximately 3 thousand year ago. In other words at about the time that the story of Moses and Aaron is thought to have taken place.

Evidence From the Lemba— The Black Jews of Africa

Additional evidence for the same unique modification of DNA among Jews comes from a study of the Lemba, a Southern African tribe. Analysis of changes in the Y chromosome show that the Lemba can be subdivided into those of Bantu and those of Semitic origin. This fits with Lemba oral tradition which describes an emigration from somewhere in the north, at least 1000 years ago. Moreover among the Lemba of apparently Semitic original there is a subgroup of priests who have inherited the Jewish Bible and attempt to promulgate its teaching. These priests consider themselves "Cohens" and it turns out that their Y chromosomes support this belief. The genetic signature of the Cohens is found among the Lemba priests in the same relatively high proportion as that found in Caucasian Cohens. The conservation of this genetic signature among males is thought to be due to the tribal custom of forbidding marriages to men from outside the tribe.

The reader may have noticed that in beginning our discussion of the Lemba we indicated that something about the Y chromosome unites the Semitic Lemba to Semites elsewhere in the World. Indeed some genetic alterations on the Y chromosome are shared generally by both Ashkenazi and Sephardic Jews, as well as the Semitic Lemba. The distribution of genetic changes within the Y chromosome distinguishes Jews from non-Jews throughout Europe.

Genetic Similarities Among All Semites

Genetic signatures have been obtained which not only unite Jews from diverse countries and continents but also unite Jewish with non Jewish Semites, that is to say with Arabs. Since Jew and Arabs were present in the same land, millennia ago, this may not seem surprising. However one must remember that the Jews showed a great resistance to marriage out of their group. Such marriages did of course take place as especially memorialized in the Biblical Book of Ruth. Moreover, since Moses' father-in-law Jethro was not from the 12 tribes we can conclude that Mose's wife was not a Jew. Nevertheless the prohibitions against out of clan marriage would suggest that the common genetic signature of today's Jews and Arabs probably does not exist because of marriage between the two. It is more likely that the genetic signature exists because both peoples spring from a common set of ancestors predating the Mosaic period. This conclusion is of course, the one supported by the Biblical story of Abraham's other "wife" or concubine, Haggar. Haggar bears Abraham his first son, Ishmael. He is ultimately sent into exile because Abraham's wife Sarah is jealous and fearful that her own child, Abraham's second son, Isaac, will lose his patrimony. God promises both descendants of Abraham that each will give rise to a great nation. Studies of the genetic signature which relates Semites to one another, suggest that that signature originated 1000 or more years before the genetic signature that relate Cohens to one another. In other words, today's genomes preserve the sequence described in the Bible, which tells us that Abraham gave rise to the Semitic lineage long before the Mosaic figure appears on the Biblical scene.

Curiously, among Moslem Arabs living in Israel and Palestine there is an unusually high proportion of a Y chromosome subtype found in Jewish Cohens—part of the "priestly" signature. This disturbs the apparent concordance of genetics with the text of the first two Biblical Books, Genesis and Exodus, because it implies that the "priestly" lineage predates the stories of Moses and Aaron, and already existed during the time of Abraham. However the "windows" of time provided by the genetic analyses are very large. The analyses of the "priestly" signature indicates that it is at least 3000 years old but it could be as old as four or five thousand years. Hence, the priestly genetic signature could have originated in Abraham's time and the high proportion of "priestly" genes among Israeli and Palestinian Arabs could be consistent with the stories of Abraham, Sarah, and their Jewish descendants, and with the story of Abraham, his concubine Haggar, their son Ishmael and his Arab descendants. If so, then something is missing from the

Biblical narrative that would trace a priestly line from Abraham to Aaron and the same priestly line to Ishmael. But how could this be? If Ishmael and Abraham's son Isaac shared only a father in common and if they also shared a genetic signature that became selectively distributed among subgroups of Arabs and Jews, then why don't all of Abraham's children carry that genetic signature? Or, if they did, why was it not transmitted to all their descendants?

We cannot answer these questions. But it doesn't really matter. Indeed the lineage of Moses and Aaron is shrouded in an impenetrable Biblical haze. And the existence of Moses and Aaron as historical figures has been questioned by generations of scholars. What matters instead is the idea that there are definite genetic patterns that relate large numbers of persons of Jewish lineage and that, in agreement with Biblical texts, the Arabs and Jews are certainly genetic brethren.

That is not to say that there is a monotonously similar genetic picture among Jews or among Jews and Arabs. For example 32% of Arabs have one pattern of Y chromosome changes that is rarely seen among Jews. Moreover within various regions of the Middle East there are differences in the pattern of Y-chromosome changes between geographical subsets of Arabs and between subsets of Jews. These differences indicate admixture between the various subsets and their surrounding non-Jewish or non-Arab neighbors.

Jewish Genes—Ethics Of The Torah

None of the genetic changes involved in the preceding cases have been associated with either physical or psychological traits. But we can now accept the existence of pieces of DNA that travel preferentially with Jews. We can even point to pieces of DNA that travel with specific subsets of Jews thought to be descendants of special, priestly or holy figures. If this is so, then we may certainly believe that other pieces of DNA permitted selected ancient figures to develop ideas that distinguished them from the common run of humans. And we can believe that some selected pieces of DNA enabled a group of surrounding persons to adopt, at least in principle, the religious and ethical theory promulgated by the leaders.

However, as pointed out in an earlier chapter, a genetic explanation for the prophetic message does not enable us to decide among the alternatives presented for the ways in which the prophets experienced their receipt of the message. Did it come to them unaided by drugs, fasting or thirst? Did it appear outside the context of mental illness? Could the prophets have made up their experiences and claimed that they came from God or through the voice of God, in order to have the greatest possible influence over an audience that was highly superstitious and had always believed in the intervention of supernatural figures? Such a possibility— pretending for the sake of the greater good— does not mean that the pretenders lacked a belief in God. They could certainly have been believers who wanted to use the force of God to further their own ethical aims. Chapter eight discusses this possibility. But before we get there we ought to ask ourselves what purpose underlies all this energy—genetic or otherwise— directed toward producing thoughts and behaviors that we now call "good"? Why be good?—that is the subject of the next chapter.

CHAPTER SEVEN

WHY ARE PEOPLE GOOD?

The ethical precepts promulgated by Judaism and perhaps first imported in large part from Egypt are a blueprint for what may be called the "good" in people. The Hebrew scriptures tell us to be concerned about the welfare of our fellow human beings. The message was presented to the people by Moses and the prophets. I have suggested that leaders like the prophets were unique. I have suggested that their acceptance of this code and their decision to preach it was driven by their genes. I also suggest that genes permitted the audience to whom the prophets spoke to adopt the new message, at least as a goal toward which they could strive. But why should they strive in that direction? Of what practical use was the new message? The question "why be good?" has resonated through the ages. The search for the answer to that question has been central to the efforts of philosophers for over two thousand years.

Fear of Punishment—Of The Individual Or Of the Society On Earth Or By The Loss Of Heaven?

During the last two millennia those philosophers who believed in the existence of a divine being or force ascribed what goodness they could find in mankind to the adaptation of rules presented to mankind by the Deity. These moral philosophers, in a predominantly Christian society, generally related the willingness of people to obey God's laws to the fear that those same people had of God's punishment. Christian doctrine presented the absence of a heavenly hereafter or the absence of a bodily resurrection as ultimate punishment

Judaism, however, has never emphasized the divine punishment of individuals as the reason for good behavior. In the last centuries prior to the birth of Jesus, Judaism certainly incorporated into its literature concerns about a bodily resurrection, about the fate of the soul, and about the existence of a heavenly realm. But Jewish writings, including the Bible, much more frequently and vividly spoke of punishment directed toward the entire Jewish people. This punishment was God's desertion of the Jewish people, ultimately manifest in their exile and in destruction of the Temple and of Jerusalem. The primary sin was expressed in the Bible as the worship of other Gods. I should like to suggest that this sin is simply a metaphor used by the ethical geniuses who founded Judaism. This metaphor refers to the abandoning of the moral prescriptions presented in the Biblical text. The punishment refers to societal disintegration which is an ever present danger to a people in exile. By being "good" we avoid the decay of society; by being "good" we bring benefits to this World not only for our own sake and for the sake of our descendants but also for the benefit of all who surround us. Hebrew scriptures ask us to do this without regard to rewards in a future life. In return the benefits will be reaped by those on Earth; heaven will take care of itself.

What Kind of Earthly Benefits Do The "Good" Reap? The Greek Perspective

Some Greek philosophers shared the perception that earthly benefits will follow good behavior. For Plato and Aristotle these benefits resided in the stability of the State. They urged their listeners to behave well in the interests of achieving a psychological condition that has been most commonly translated as "happiness". They asserted that in a perfect society this "happiness" would exist when behavior meshed with the best interests of the State. They also recommended a much misunderstood principal of the "mean". This principle tells us that in all things we should act or feel in a way that is appropriate to the matter at hand; that is neither over reacting nor under reacting. Contrary to what many people have suggested, this does not mean that one must always seek a response in the middle of the road. Rather the rule of the mean should lead to a greater or a lesser response, depending upon what is really appropriate under the circumstances. The more perfectly disciplined the person, the more likely he or she will accurately perceive what feeling or response is dictated by the rule of the mean. Such a person is governed less by passion than by reason.

But what the ancient Greeks called ethics are really personality traits like courage or behavior related to support of the interests of the Nation or State. For these Greek philosophers the highest professions, aside from that of philosopher itself, were those of military leader or politician. How different from the Jewish view, illustrated in the Biblical portrait of their first King, Saul. He is so unwilling to serve that he hides from his annointers.

Greek writings about ethics are curiously devoid of what we would consider moral dictums. We do not find proscriptions against murder or adultery. We do not find rules for the treatment of widows and orphans. The closest we come are the traits of liberality and of justice. The former includes giving to others in accordance with the principal of the mean. Justice includes punishment for transgressions. In a list of the latter we find murder, adultery, theft, false witness, slander, robbery. These are among the evils specifically proscribed in the Bible. Aristotle's "Ethics" deals with them in the rather oblique manner just described; a list of transgressions appearing in a section devoted to justice and punishment. Perhaps as a result of such tangential treatment of ethics in the works of the Greek philosophers some scholars have asserted that the great Greek writings on ethics give us no clue about the morality of the common man in the Greek empire or city-states. Recognition of this gap in knowledge has led scholars to search for hints of everyday Greek morality in the plays and poems of the period. But these analyses also provide little insight into Greek thought about the moral issues at the epicenter of Judaic laws; that is, concern for the welfare of one's neighbor and for the poor, the widowed and the orphan or for what we may call social justice in general.

Biblical Ethics Precede The
Golden Age of Greece

Moreover even if the Greek writings did contain the same moral concerns as the Bible, we would still ask whether this invalidates our claim that Jewish thought was unique to its time and place. These Greeks introduced their philosophy in the middle and latter half of the millennium preceding the birth of Jesus. This period in ancient Greece followed by at least 500 years, the time traditionally assigned to Moses, and by at least two hundred years, the historic reign of King Josiah of Judah; a reign in which the Biblical Book of Kings tells us that the Mosaic laws were rediscovered. If the Greeks did share the ethical beliefs of the Jews it is puzzling that they failed to acknowledge a debt to Judaism.

In fact, Herodotus, the Greek who some consider the World's first historian, made no mention of the Jews even though he described in great detail all of Earth's peoples and much of its fauna as they were known to him. Why were the Jews omitted? Other ancient records have been discovered which note the military strength and accomplishments of the Kingdom of Israel. These events occurred prior to the flowering of ancient Greece and surely should have been known by Herodotus. The mystery of Herodotus' silence is even greater when one reflects on the geographical proximity of Greece to the land of Israel. Perhaps the omission was deliberate and reflects the great anti-Semitism of the Greeks. Josephus, a Roman-Jewish historian, responded to this antisemitism in a polemic titled "Against Apion", written shortly after Jesus death. There Josephus attacks the anti-Semitic claims of the Greeks; claims like the assertion that Jews did not partake in a miraculous exodus but instead were lepers expelled from Egypt. In any case, the ethical writings of the Greeks do not precede Jewish ethical concerns nor do they develop in parallel with them. Rather the Jewish ideas appear first and are much more focused upon the treatment of one's neighbor.

Why Are Humans Moved By
Concerns For Societal Good

The Greeks, however, do appear to define "good" behavior in terms of what is best for society. In this they appear more explicit than the Bible, in which we find the same rationale only after interpreting God's punishment—withdrawal from the people—as a metaphor for dissolution of the State. But why should societal good become the standard by which individual morality is judged? Why doesn't what is good for me take precedence over what is good for all of us? From where do we get the notion that what is best for all of us is best for me? Why shouldn't self preservation, even when it is obviously at the expense of society, come before concerns for the survival of the society?

Neither Jewish nor Greek thought provide an answer. The Greeks suggest that a society at peace with itself will provide its citizens, and especially its privileged philosopher-kings, the milieu in which they can develop the highest knowledge. With this knowledge they can thus approach "truth". "Truth" they say, may be the greatest "beauty". Perhaps this greatest or perfect beauty is simply another metaphor for the Judeo-Christian-Islamic God. Then the pursuit of God—or truth and the most perfect beauty— must, according to the Greeks, lead to preservation of society. The ancient Jews also sought God in order to preserve their society. If they deserted God's precepts then God would desert them.

The nature of the society praised by the Greeks certainly differed from that sought by the Jews. Many of the details in both societal ideals differ considerably from today's Western views of the ideal society. But in the final analysis it may not be the survival of their particular society that directs people toward goals that they believe are favorable to that society. A more deeply seated drive may be the drive toward survival of the human race; the human species.

The Genetic Drive To Preserve Our Species

But why do we strive, even unconsciously, for preservation of our species? In the twentieth century, genes provided the answer. Genes permit us to be good-to act in accordance with Judaic principles. These genes can also be perceived as genes that help the species to survive. It is preservation of the human species that is the ultimate reason for being "good". On the other hand there are genes and their polymorphisms whose influence will lead to antisocial behaviors. The history of human beings is the result of these opposing influences. We also acknowledge that there is, at times, a genetic drive to preserve our individual selves. Sometimes this is appropriate But at other times the drive to protect ones individual body may conflict with the need to arrive at resolutions of conflict that are more likely to assist in the survival of humans everywhere. At no time in history has this been more evident than in the age of atomic weapons and of weapons that can deliver deadly germs over wide areas of the inhabited earth. Today we can say that it will be the expression of certain sets of genes that will lead to preservation of society. We hope that the influence of these genes will prevail.

The moral precepts that will help us to survive were announced by people who knew nothing of genes or science. They tell us that these precepts came from God. But all theologians agree that God is unknowable. Therefore the persons represented by Moses and the prophets in the Biblical account, and those who codified the account in Scripture, were substituting one unknown, God, for another entity of which they were equally ignorant and had not yet imagined. That entity is the human genome. What we must now ask is whether the prophets really believed that the precepts came from God, whom they could not have claimed to know or understand? Or we may ask, instead, were the prophets and the leaders who followed them simply pretending that their message came from God in order to bring about change in a superstitious society? The next chapter addresses this question.

CHAPTER EIGHT

PRETENDING FOR THE SAKE OF FAITH

In an earlier chapter I considered the possibilities that dreams, self induced ecstasy or mental illness were the alternatives to actual messages from and/or a visit from God. These possibilities imply that the prophets really did not see or hear God but were listening, instead, to their own, gene driven, inner voices. However, as pointed out earlier, this does not mean that they were atheists. They may even have believed that God visited them during these experiences. We can never know. But we will now examine another possibility. Did the prophets pretend to have their voices and visions in order to influence a superstitious audience to accept the Judaic vision of a better World?

If we wish to be extremely skeptical, even cynical, we might suggest that the words recorded in the earlier writings did not even exist in the historic setting provided by their texts. We could even provide later writers with a motive for falsely representing prophetic exhortations and warnings as as if they were uttered by men who lived in earlier centuries. The later writers might have made up these accounts in order to provide a retrospective explanation for the loss of the ten tribes in the Kingdom of Israel, for the conquest of the remaining tribes in the Kingdom of Judah and for the destruction of the temple and the exile to Babylon. The Jewish people had been left in a pitiful, demoralized state which clearly required some explanation if they were to retain their identity and allegiance to the God that was the focal point of that identity. Divine punishment for straying from the ethical path and the promise, even then, of a return to paths of glory in exchange for obedience to the law, certainly met the need to explain defeat and to maintain faith. Let us look at **a** basis for this suggestion in the Bible itself.

Was Deuteronomy A "Lost" And "Rediscovered" Book Or A Newly Written Book Of Propaganda?

The Book of Kings tells us that a lost version of the Law was discovered in the temple during the reign of King Josiah. The book is brought to the king's attention by the priests who found it. He then decrees that it be read in public then and each year thereafter. Through it Josiah attempts to restore his subjects to ways of righteousness. Scholars believe that the lost book is the book we now call Deuteronomy, the last of the Books of Moses. The cynic can suggest that the book was not lost and then found, but that it was actually written during Josiah's reign for the purpose of restoring the influence of the priests and returning the people to path of the Lord. Some scholars have made this suggestion.

Since Deuteronomy and its laws are largely a recapitulation of earlier portions of the Bible why would a discovery or a writing of Deuteronomy during Josiah's reign be so important? Why didn't Josiah just order a reading of the books that supposedly preceded it? Perhaps the farewell speech of Moses that is contained in Deuteronomy was essential for its charismatic poetry. More likely is the possibility that the differences between it and its predecessors, differences that may seem slight to many readers, had significance to the priests of that time, so that they wanted to supplant an earlier version of the religion with the Deuteronomic version. In any case, if the 'rediscovered" scrolls had really just been written, the account of its discovery told to the people by the priests and the king would be a false one presented for a political purpose.

Indeed some suggest that even the later Book of Kings is a revisionist account of 500 years of Jewish history from Solomon's reign to the destruction of the temple. These scholars suggest that the revisionist history was an attempt by those in the Kingdom of Judah, to glorify their Kingdom and its Davidic ancestry at the expense of Israel, the once powerful, famous, but now defunct, Northern Kingdom.

Are these suggestions reasonable? We have no smoking gun—no document that admits to a fictitious representation of the manner in which the prophets got their ideas; no document that admits to a falsification of the history in parts of the Book of Kings. However in pursuing the suggestion that one or more of the prophets may have been pretending to hear or see God, I believe that it is reasonable to look for supporting evidence in the writings that do survive from the later centuries and indeed from the fifteen hundred years or so that follow the codification of the Hebrew scriptures.

Looking At The Post Biblical Period

I believe that it is valid to consider this material because we have no reason to believe that there was an abrupt change in the genome or its expression in religious leaders arising shortly after the birth of Jesus. Nor is there reason to suppose that these leaders responded differently to their environment than did religious leaders of the previous several centuries. What they said and what they chose to record privately or for only a learned few to read, may have been governed by the politics of the societies in which they worked, but there is no reason to believe that their inner most beliefs differed from those of their predecessors. Nor is there evidence of such a change in the 1500 years that followed, either among the religious leaders or among the faithful believers whom they influenced. What did change was the recording and saving of the written record that reflects their thoughts. Therefore we can look at the statements of the undoubtedly faithful believers who lived in those later centuries. We can learn what they believed to be true and untrue. We can see whether they always spoke the truth. If not, we can see how they reconciled untruths spoken in the name of their religion with their own deep faith. We can use these records as guides to our understanding of what may have occurred in the minds of the religious leaders in the Hebrew scriptures.

Of course one might wonder whether the words of post Biblical religious leaders and of famous persons of religious faith have come down to us through a process that has preserved only the words of "winners"; those whose thoughts were in tune with the advances in knowledge that occurred in more modern times. Such persons might not have been at all like the persons who were able to assume leadership roles in a more "primitive" society. That is why I will end my period of inquiry at about the middle of the 1600's—the mid seventeenth century. I end it there because we can agree that up until this time the basic facts of nature were still a mystery to everyone. There was no understanding of bacteria or viruses and thus no understanding of the causes of disease. There were was no understanding of the causes of hurricanes, droughts, earthquakes or volcanic eruptions. The nature of the universe, the relationship of the planets to our sun, and the understanding of earth's true extent had not been glimpsed until the middle ages ended. Thus neither the leaders nor the common folk were free of the ignorance that breeds superstition and fear. Therefore, up until that time we cannot claim that new knowledge had banished superstition and had brought forth a new kind of religious leader necessitated by new times. We cannot claim that the moral leadership could have shifted from those with a genome suited to a more primitive era and had become centered in another kind of human, one whose genome was more suited to modernity. Modernity with respect to the natural and physical sciences had not yet arrived. We will look at this period of 1600 years by beginning with its end and gradually move backward toward the early Christian period.

Galileo And His Daughter

We begin with the correspondence of a daughter and a famous father dating from the beginning of the seventeenth century. Scientific enlightenment had just begun. The daughter writes the father that neither the love of ones children, nor pleasures, honors or riches can bring about true contentment because all these things are transitory. Instead she tells him that only in God, as in our final destination, can we find real peace. When death has torn away all of the ephemera, great joy will be ours because we will celebrate in God's glory, face to face with Him. The same writer, a Catholic nun, wrote in another letter to the same father that life itself is like a brief and dark winter after which God grants us the happiness of an eternal spring.

From these words there can be no doubt about the deep faith of this nun, known as Sister Maria Celeste, a daughter of Galileo. But, from other letters to her famous father, one must also conclude that she never doubted the correctness of his view that the earth and all the planets circled the sun. She now knew that the sun and not the earth was the center of our universe. She now knew that when Joshua fought the battle of Jerico it could not have been the sun which stood still as the Hebrew scriptures stated. The sun was not revolving around the Earth in the first place.

This challenge to the literal meaning of the Biblical text caused Galileo to be charged with heresy. And what of Galileo himself? Was he too a devout Catholic, a man of faith, in spite of his belief in the conclusions he drew from his study of the heavens? The answer is a resounding yes. Galileo was chastised by an inquisitorial commission. His book was banned. He escaped physical punishment and was ultimately allowed to return home. He achieved this result by telling a series of lies. He declared that his proposals were only a theory. He said that by presenting this theory he was proving to Northern Protestants that Italian Catholics were also cognizant of the new observations of the heavens. But Galileo said that the theory he presented was an incorrect interpretation of those facts. By presenting the facts and the theory he said he hoped to prevent others from falling into error and accepting the false theory of a sun centered universe! He pretended to be performing a public service so that the flaws in the theory could be found and the theory demolished. Galileo told his inquisitors that he believed his book clearly showed the errors of scientific reformers—the weaknesses in their reasoning. He also said that when he had been warned earlier to abandon this theory, which had first been expressed by Copernicus in Poland, he, Galileo, immediately abandoned it as instructed. Galileo said, "I have not held this opinion after [I was instructed to abandon it]". When his examiners doubted the truth of such statements and condemned him anyway, he told his examiners that he disavowed ". . . with a sincere heart and unfeigned faith" any errors that had entered his writings, and indeed any error "contrary to the Holy Catholic Church".

Galileo's disavowal of his belief was certainly a lie. But was it told only to save his skin? That is a commonly held assertion. I think the assertion may be incorrect. Surely he realized that he must tell the lie in order to save himself from torture and perhaps death. But he was also the devout Catholic he painted himself to be; a Catholic just as devout as his daughter. He understood that leading Churchmen believed sincerely that the underpinnings of the Church were threatened when the literal truth of Biblical passages were called into doubt. Galileo believed, as the 20th century Church has stated, that there is no contradiction between faith and science; or as he would put it—between faith and nature. He believed that all discoveries of the natural order were simply unveilings of Gods handiwork. Biblical passages that apparently contradicted the new discoveries had to be reinterpreted. Perhaps the Bible meant only to say that the sun appeared to stand still. Indeed, pointing to the passage which said that the sun remained in the center of the sky he showed how that observation would be most understandable if it was the Earth rather than the sun which had actually stopped moving. He believed in the Holiness of Scripture, if not in its interpretation. He recanted his belief in his own conclusions about nature, not only to save himself, but to help the Church maintain its authority. He was told by the churchmen he venerated that the authority of the church depended upon literal acceptance of the words in the Bible. This, they felt, was especially required in the aftermath of the Protestant Reformation which appeared to leave it up to every person to interpret the Bible in his or her own way.

If we are too cynical to accept this more nuanced interpretation of Galileo's recantation, then what are we to make of his devout daughter, who served as her father's copyist? She not only read and rewrote his words in her own fair hand but also is thought to have understood them. Why, in her letters to him, is there never a word from her, chastising him for a heretical belief, or for denying the literal truth of Scripture? Did she too believe that one had only to reinterpret the Biblical passages in question, and that therefore her father's proofs were not, in fact, heretical? Or did she, who was never approached by the inquisition, agree silently to accept its instructions to her father and to the Catholic community at large? Did she decide to disavow, even to herself, the truth of her father's conclusions in order to preserve the integrity of the Church whose faith was clearly the basis of her spiritual existence? Or did she believe that the biblical passage was simply a legend that her father had proven false? Did she then agree to accept the falsehood in the name of the Holy Church? At the very least, in view of her failure to challenge her father for religious error, it seems reasonable to believe that she too recognized that the error lay instead in the literal interpretation of the scriptures. Her silence is one way to hide one's understanding of the truth in order to maintain the authority of religious leaders.

Martin Luther's Crisis Of Faith

If we move back in time just a few years we come to Martin Luther. Luther was the origin of the Church's fears concerning new interpretations of scripture. Luther encouraged every person to examine the Bible for themselves. His German translation made this possible in much of central Europe. No one can doubt Luther's faith. I present him for the readers consideration as a man of deep faith who, if he did not deliberately lie when he preached, at least wondered if what he was saying was true. This possibility is presented in a recent, scholarly biography of Luther in which the author examines many of the statements Luther made to his acolytes as they sat with him at meals. Day after day, and month after month, they recorded them in voluminous journals preserved to this very day.

To many earlier scholars, Luther's greatest fear has been described as a fear of the Devil; a fear of damnation for his sins. But examination of his statements fails to provide much support for that point of view. Rather, what emerges is grave concern about the truthfulness of a basic doctrine: is there resurrection even for those justified by faith? Is it possible that death really is the end of everything? If so, then life is really more than one can bare; a wicked trick played by God for no purpose. This, according to his own statements, was the fear that tortured Luther for much of his life. If that is so, then we certainly have the case of a devout believer in a real God, who preached a sometimes consoling doctrine of afterlife in heaven, at least for some of us, but who continually lived in mortal fear that this was untrue, not only for himself but for many persons of faith.

Moreover, Luther questioned the Holy origin of much of accepted scripture. He suggested, much as I have suggested here, that great religious leaders may have bent the truth when speaking to common men and women in order to achieve a greater moral good. The apostle Paul, the true founder of the Christian faith speaks with certainty about the crown of life laid up for him in heaven. Luther is recorded as doubting that Paul believed as strongly as he spoke. Yet for Luther, it was Paul who served as the Christian beacon.

Was Unbelief Possible When The Middle Ages Ended?

The French historian, Lucien Febvre has dealt with precisely this problem—was it possible to lack faith in an era when faith was the apparent backbone of everyone's existence? He writes, in the twentieth century, a book titled "The Problem Of Unbelief In The 16th Century". In this book, Febvre insists that such unbelief was not possible. If he is correct, then at least in the sixteenth century it would not have been possible for persons of faith to actually disbelieve what they were preaching. If the mind of humans then was essentially the same as that 2000 years earlier, then those earlier teachers of Jewish ethics must also have believed all that they told there listeners. But is Febvre's analysis correct? To answer that question we must ask what he meant by "unbelief"

Febvre's book was written as a response to the work of Lefranc, another French historian. Febvre tells us that Lefranc has accused Rabelais, France's most famous sixteenth century writer, of being an atheist. Febvre counters by insisting that no Frenchman could have been an atheist during the sixteenth century. But a careful reading of his book informs us that what Lefranc has really done is accuse of Rabelais of not being a Christian. That, for Febvre, a French Catholic, is the same as an accusation of atheism. Such a definition of atheism was and is, of course, erroneous. Nevertheless Febvre's book is extremely useful to us, because it brilliantly focuses on questions that concern our own quest to understand the origins and acceptance of Judaic ethics.

Febvre discusses the effect of philosophy and Aristotelian science or logic upon religious thought. He tells us that for more than a millennium, there were always at least some important leaders of the Christian community who spoke of the Bible as allegory. He equates the thinking of Rabelais with the thought of Erasmus, a great contemporary of Luther. Erasmus never denied his Catholic faith. He refused to join the Protestant side. Yet many Catholics were suspicious of his beliefs because he asserted that both Old and New Testaments might contain allegory rather than literal truth. Febvre does not share that suspicion. He tells us that both Erasmus and Rabelais were men of faith.

Febvre defends Rabelais against the charge of atheism by asserting that Rabelais must surely have been a Christian, whatever attempts he may have made to square Biblical text with the realities of nature. We might say that for Febvre, Rabelais was but another Galileo. Indeed Febvre tells us that for some scholarly Christians there had always existed a frame of mind which said that philosophy was one thing and religion another. These men said that ". . . for all Christians the doctrine of Christ must prevail", regardless of what might appear as philosophical truths or the truths arrived at through scientific observation. And Febvre allows that some people made such statements sincerely and were not dissembling. But this must mean that in an age of deep Christian faith, some people must have been capable of holding within their minds two apparently contradictory beliefs while at the same time denying the contradiction. This point of view is exactly that which I have applied to Galileo and to his daughter.

But one of two contradictory beliefs must be false. If devout persons can divide their minds in this way, for the good of the church, then surely we can conclude that one half of the mind was falsifying the truth to the other.

The Europe of the sixteenth century was peopled by extremely devout persons in all classes of life. In such an atmosphere would it have been possible to be an atheist while at the same time teaching others that the ethical code promoted by religion was a code coming from God? Febvre insists that the mind set of sixteenth France precluded true atheism. As I said earlier he defines atheism as the absence of Christian belief. The existence of Jews or Muslims simply does not appear to strike him as a contradiction to his opinion. But, even for Christians, Febvre's own words ultimately fail to pass the test of logic. He carefully spells out the Christian dogma, belief in which is synonymous with his definition of a Christian. But then he writes in a chapter titled "A Century That Wanted to Believe", that an allegorical interpretation of the Bible is permissible and not equivalent to atheism. "Let us be fair" he says to the men of the sixteenth century, who were "attempting to restore unity of thought" and to establish "harmony between . . . the facts of nature and their conception of the Deity". If Febvre concedes that much, then I am entitled to ask, did such men really believe that God spoke to the prophets with a human voice and that God presented itself to the prophets within what they reported to be visions? Moreover, if I may ask this question about the religious people of the sixteenth century then I believe that I can ask it of the prophets themselves. Didn't such men speak in allegorical terms without admitting that to their audience? Didn't they do so in order to promote the moral principles of the Judaic World view by linking it to the authority of the Deity as the source of moral instruction and as the provider of reward or the provider of punishment for obedience or disobedience to the new code? I suggest that this is exactly what happened.

Augustine And Origen

Let us leap backward more than a thousand years from the sixteenth century. Among the great leaders of the early church were Augustine and his predecessor Origen. Both men spoke of allegorical passages in the Bible. For example, in Book 12, chapter 24, of his monumental work "The City of God", Augustine writes that God did not take the rib from Adam in a material way, but rather in a divine way. The reader is cautioned against believing that God works as an ordinary artist molding materials into human shape with His hands. Augustine chastises those who label as fables, stories like that of the creation of Eve. They are only fables, he says, if you take them literally, applying conventional standards of craftsmanship to the work of God. In Book 11, chapter 21, Augustine tells us that the mind of God should not be thought of as if it were a human mind. Indeed God's mind is not a mind at all in any human sense.

Augustine's assertion that portions of the Bible are allegory is mixed with assertions that other portions of the Bible are literally true, no matter how contrary they are to the experience of humans living in his (and our) own time. Thus he insists that the long lifespan of the earliest Biblical characters, including Methuselah's 900 years, are absolutely correct. Such mixtures of belief and disbelief in the literal truth of the Bible, on the part of so devout a Christian, support our earlier statements concerning the ability of believers to hold opinions that would be at odds with those of persons claiming absolute, literal devotion to the Biblical text. But Augustine did not write his book for the masses. It was written in Latin and could only be read by the literate whose number was exceedingly small in his time. Indeed the "City of God" was written at the behest of leading Christian Romans at a time when Rome had suffered great calamities at the hands of invading marauders. The Empire was well into its fall. Pagans blamed it all on the spread of Christianity and on the parallel desertion of the Pagan Gods who, they believed, had protected Rome. Augustine's friends begged him to refute these charges. The book was to be read by the society's leaders, Pagan or Christian, and not by the person in the street.

To the latter, Augustine was certainly not above preaching as if the Bible were literally true in all its respects while at the same time he believed that this was not, in fact, the case. Indeed, on page 293 of his "Confessions", he tells us that the words of the Hebrew Scriptures, as set down by Moses, were chosen to be comprehensible to their audience. For example, on page 282 of the same work, he tells us that in Genesis the words "earth visible and without form" were meant to convey to humans "in a way that they can understand, that formless matter which You created without beauty in order to make from it this beautiful world". Similarly, on page 114 he writes that he understands that God really did not make man "in His own image", but that this too is simply a phrase placed in the text to give the human listener some sense of his own immense potential. In Book Seven of the "Confessions", Augustine tells us that he became a Christian only when he was able to understand that God did not have a human image, but rather was without bodily substance. In chapter 27 of Book Twelve he goes on to write that there are people who think of God as actually speaking the words "let there be light, etc." and that what was ordered came immediately into being. About such people—one may assume the general mass of religious people at the time—Augustine writes that ". . . they are like children . . . . the simplicity of the language of Scripture sustains them in their weakness . . . . on [this simple language] is built their faith". Clearly then, one must preach to these masses in the language that they understand. Augustine does not make fun of these common people. Rather he chastises others who would consider the words of the Bible merely as words fit for "simpletons". Instead in chapters 28 and 29 of Book Twelve he glories in the words which lend themselves to a multitude of interpretations by persons with a disposition to do so. He tells us that for such persons the words of the Bible are a "leafy orchard" in which one may find "hidden fruit". Provided that readers do not stray too far from what he believes, Augustine welcomes diversity of interpretive opinion. "There is truth in each of them".

Augustine wrote in the fourth century after the birth of Jesus. Origen wrote one hundred years earlier. Scholars consider some of his writings to be the first significant attempt of Christianity to present a systematic theology. In spite of this, some of his work was considered heretical and was condemned. Among these were his comments about damnation. Origen doubted that the flames of Hell were real; a point much debated by early and later Christian fathers. But the flames of Hell are not part of the Hebrew scriptures. Origen was referring to early Christian imagery, much disputed by later authorities. In any case Origen was clearly willing to consider standard theological imagery as allegory or metaphor rather than as literal text. This insistence on allegorical interpretation of the Bible may again have been aimed at those who could read—in this case readers of Greek. One wonders whether he would have modified his preaching and dissembled with common folk in his audience, by pretending that the text was literally true.

On the other hand, Origen was also capable of insisting that certain Biblical passages were neither literally true nor allegory in a poetic sense, but rather that they were coded forms of a hidden truth. In such cases the words alone simply did not communicate the true meaning of the text. This is apparent in the writings he directed toward other theologians. His work in this area is among the earlier Christian efforts at interpreting portions of the Hebrew scriptures as texts that foretold or actually represented events which Christians believed occurred later in history. For example in a text about the Jewish Passover festival, Origen tells his readers that the true meaning of the Passover is the "passing over" of the faithful onto Christ. And, within this big picture, he reinterprets the details, telling us, for example, that the sacrificial lamb is, of course, Jesus himself. Here we have, as in Augustine, that curiously illogical gift of men of deep faith, which permits them to accept the words of the Bible as the words of God, but permits them, at the same time to believe that selected words or passages are not, in fact, the true word or the plain meaning of Him who instructed Moses. Rather the words are code whose true meaning should be revealed to the masses by the initiated.

We have now seen that the earliest Church Fathers were fully capable of believing that portions of the Hebrew Scriptures were allegory or metaphor for the events that they described or were code describing future events. We have seen that as early as Augustine, founders of the edifice now known as the Catholic Church, understood that the words of the Hebrew Scriptures might be treated literally where untutored, simple people were being taught, and that the use of those words, by their author, provided a means of communication with those who would otherwise be unable to grasp the essential message and teaching of the text. The adaptation of the moral teachings as a way of life depended upon the belief of superstitious people in the fact that they came from a powerful and caring God. Therefore, one had to use words which reinforced that belief among the members of the audience.

If the early Churchmen understood these things, why should we doubt that the founders of the Jewish religion, whose moral teachings the Church adapted, also understood that the words in the Scriptures might often be allegory or metaphor? Why doubt that even the prophets themselves understood that unsophisticated listeners had to be influenced by grand words and imagery which presented a picture of a God who spoke and looked like them and who had a mind that reasoned and even admitted to error. If these ethical leaders were believers in that supernatural concept called God, that belief need not have meant literal acceptance of all that they taught, any more than it did to the minds of Origen or Augustine, or to the minds of churchmen and women in France and Italy during the Reformation and the two century that followed.

There are many ways to mislead a listener. One can say things that one knows to be false. One can appear through silence, to be in agreement with statements one knows to be false. One can tell only part of a story, leaving out elements essential to a correct understanding of the described events. I suggest that the prophets misled their audiences in one or more of these ways. I suggest that the prophets felt the call of their own genes. I suggest that these genes and not the literal voice of God, nor images of God or of angels, led the prophets to conclude that the general welfare would benefit by adapting the ethical code that they preached. If their images and voices came to them in dreams or through the use of drugs or during self induced episodes of deprivation I suggest that the nature of the messages they received was still dependent upon their genes as was the desire to pass the message on. Moreover, even then, it is entirely possible that they knew that these experiences were hallucinatory in nature and simply used them as motivational devices to assist them in what was an extraordinarily difficult and often dangerous career.

EPILOGUE

WHERE DOES THIS LEAVE GOD?

I have suggested that the prophets and the code of Judaic ethics that they preached revolutionized Western ethics. Though much may have been imported from Egypt, the Judaic adaptation restored the concept of one God, made that God central to the existence of an identifiable people, presented that people with a God given set of rituals and an extensive set of ethical proscriptions. Most importantly the prophets made clear that obedience to the latter was as essential [more essential] to that people's survival as was the slavish following of the rituals. I have suggested that these teachings, the propensity to preach them and the ability of the community to accept them as a behavioral goal, are all the result of genetic change. What about God?

None of these assertions require the abandonment of belief in God. They do demand disbelief in the literal texts which say that God spoke or that humans heard God's voice or saw angels or other representatives of God's presence. Both believers and nonbelievers in God can accept the thesis presented here. For the nonbeliever, the prophets must have been untruthful because God does not exist. The believer may resolve the issue in several ways. God may have "inspired" humans to see the World through the prism of Judaic ethics. This could have happened without the literal speech of God or angels. God may be responsible for our genes and thus God's will does not require that God actually spoke to humans or that the prophets actually saw the images that they claimed to see or were transported by angels through time and space.

In any case, it was not my intention to analyze the nature of the reader's beliefs. It was my intention to focus on the ethical leaders represented in the Bible. Although I have suggested that these figures did not actually hear the voice of God, this does not mean that they did not believe in God or in supernatural phenomena. They probably did believe in both. They probably shared this belief with their audience. What I assert is that they made use of this belief system to spread the new ethics; they pretended to hear God. I suggest that such pretense is not such a bad thing. I presented well documented examples of statements from men of great faith, even great religious leaders, who, in equally superstitious times, clearly doubted that God could literally speak or that God had human form. Nevertheless they found it necessary to use such figures of speech in order to teach ordinary humans the will of God. Their audience must have believed such stories or else the stories would not have spread. For Western civilization to move forward it was merely necessary for there to be a gene driven acceptance of the message by a critical mass of people. The message was taught by persons with an even more distinct genome which drove them to formulate and to teach the Judaic ethical code.

There are alternatives to this theory which still do not require a belief in the voice of God. I have discussed two of these theories. One says that the voices and visions were hallucinations. These could have been a product of drugs, taken to elicit such experiences as suggested in Second Esdras. They could also have been a product of long periods of fasting in the hot desert as suggested in same Book. Such environmentally induced hallucinations need not mean that belief in God was absent from those who sought them out. They may or may not have believed that God was literally encountered on such occasions. In either case, the genes are not removed from the equation. One must still ask—why these persons and not others? One must also ask why the hallucinatory messages they received took the form that they did? Again we may invoke the genes to answer these questions.

Another alternative to deliberate lying is the one called mental illness. Here the hallucinations are not a product of drugs or fasts but a manifestation of disease. This too will be dependent on the genes. I have shown by reference to the Schreber case, that severely delusional individuals may still function as community leaders and that their intellectual functioning may continue unimpaired outside the areas of their delusions. If Jeremiah, or Isaiah were delusional they may still have been capable of giving the superbly accurate political advice reported in the Book of Kings. And certainly, even if delusional, some prophets may have been brilliant orators capable of influencing masses of people.

But for me, drugs, fasting, or illness all seem to diminish the greatness of the Abrahamic figure, and of Moses and the other great teachers of Judaic ethics as described in the Hebrew Scriptures. Uninfluenced by any of these crutches or illnesses, ethical genius, driven by a unique set of genetic characteristics, seems a far more admirable explanation of the new World view. Readers must make up their own minds.

BIBLIOGRAPHY

This book is intended for the general reader with an interest in the relationship between religion and ethics. Numerous works served as influences for the thoughts expressed here. The most important are cited below. The references vary greatly in difficulty. Of particular concern to the author is the complexity of many of the references concerning genetics. The central thesis of the book is that the development and acceptance of Judaic ethics was genetically driven. Consequently it was necessary to explain genes and their relationship to behavior in general, as well as to discuss the existence of "Jewish genes". Since this book is written for non-scientists, I may be charged by scientist-readers with oversimplifying the scientific material. On the other hand, I am afraid that the nonscientist may be somewhat daunted by the cited literature pertaining to the two chapters on genes. While I tried to simplify those chapters, the literature upon which they are based is necessarily technical.

God, Ancient Cultures, The Egyptians, Hebrew Writings

ARISTOTLE. Ethics. Transl by H Rackham, Harvard U Press, Cambridge MA 1999.

ARMSTRONG K. A History of God. Ballantine Books NY 1994

BREASTED JH. The Dawn of Conscience, Scribners NY 1933

DOVER KJ. Greek Popular Morality In The Time Of Plato And Aristotle. Publ Hackett, Indianapolis 1994.

ELIADE M. Gods, Goddesses and Myths of Creation. Publ Harper and Row, NY, 1967,1974.

GOODSPEED EJ. [Transl]. The Apocrypha. Vintage Books Edition, Publ. Random House, NY 1989

FINKELSTEIN I, SILBERMAN NA. The Bible Unearthed. Publ The Free Press, NY 2001.

HERODOTUS. The Histories. Publ Penguin Classics, London 1996]

JOSEPHUS, F. The Complete Works. Publ Kregel, Grand Rapids Michigan 1981.

KING LW [Translator]. The Code of Hammurabi. In: Exploring Ancient World Cultures at http://eawc.evansville.edu/anthology/hammurabi.htm].

MILES J. God-A Biography. Publ Knopf, NY, 1995.

PLATO. The Republic. Transl by R Waterfield. Publ Oxford U Press, Oxford 1994.

PRITCHARD JB. Ed. Ancient Near Eastern Texts Relating to the Old Testament-Third Edition With Supplement . Princeton U Press, Princeton NJ 1969

SPEISER EA. Ancient Near East Texts. Publ Princeton U Press, Princeton NJ, 1950;

STONE ME. [Editor]. Jewish Writings of the Second Temple Period. Publ Van Gorcum Fortress Press, Philadelphia 1984.

TANAKH—THE HOLY SCRIPTURES. Publ Jewish Publication Society, Philadelphia, 1988.

TELUSHKIN J. Biblical Literacy. Publ Morrow, NY, 1997.

URMSON JO. Aristotle's Ethics. Publ Blackwell, Oxford 1999.

WIGODER GD [Editor]. The New Standard Jewish Encyclopedia, 7th edition,. Publ Facts On File, NY 1982.

About Genes

ANAGNOSTOUPOULOS AV, MOBRAATEN LE, SHARP JJ, DAVISSON MT. Transgenic And Knockout Databases; Behavioral Profiles of Mouse Mutants. Physiol and Behaviour 2001;73:675-689.

BOUCHARD DT JR, LYKKEN DT, MCGUE M, SEGAL NL, TELLEGEN A. Source Of Human Psychological Differences: The Minnesota Study Of Twins Reared Apart. Science 1990; 250:223-228.

CARTER TA, DEL RIO JA, GREENHALL JA, LATRONICA ML, LOCKHART DJ.BARLOW C. Chipping Away At Complex Behavior; Transcriptome/Phenotype Correlations In The Mouse Brain. Physiology And Behavior 2001; 73: 849-857.

DIMAURO S, ANDREU AL. Mutations In mtDNA: Are We Scraping the Bottom Of the barrel? Brain Pathol 2000; 10:431-441.

EHRLICH GE. Genetics of Familial Mediterranean Fever and its Implications. Annals Int Med 1998; 129:581-582.

ENCODE −Nature Collections−a supplement to Nature Publishing Group Sept 2012

HALEY RW, BILLECKE S, LADU BN. Association of low PON 1 type Q (type A) arylesterase activity with neurologic complexes in Gulf war veterans. Toxicol Appl Pharmacol 1999; 157:227-233.

HELDERMAN VAN DER ENDEN AT, MAASWINKEL-MOOIJI PD, HOOGENDOORN E, WILLEMSEN LOSEKOOT M, OOSTRA BA. Monozygotic Twin Brothers With The Fragile X Syndrome:Different CGG Repeats Effect Mental Capacities. J Med Genet 1999; 36:253-257.

KASTANGO KB, DEKOSKY ST, RERRELL RE. The 5-Httpr*S/*L Polymorphism And Aggressive Behavior In Alzheimers Disease. Arch Neurol 2001; 58; 1425-1428.

LYKKEN DT, BOUCHARD TJ JR, MCGUE M, TELLEGEN A.

Heritability Of Interests : A Twin Study. J Appl Psycho 1193;78:649-661.

MCGUFFIN P, ASHERSON P, OWEN M, FARMER A. The Strength of the Genetic Effect. Is There Room For An Environmental Influence In The Aetiology Of Schizophrenia? Br J Psychiatry 1994; 164:593-599.

MCGUFFIN P AND THAPAR A. The Genetics of Personality Disorder. Br J Psychiatry 1992; 160:12-23.

HTTP://WWW.EXIQON.COM/WHAT-ARE-MICRO-RNAs

ROSENTHAL D, WENDER PH, KETY SS, WEINER J, SCHULSINGER F. The Adopted Away Offspring Of Schizophrenics. Am J Psychiat 1971; 128: 307.

SUKONICK DL, POLLOCK BG, SWEET RA, MULSANT BH, ROSEN J, KLUNK, WE,

TELLEGEN A, LYKKEN DT, BOUCHARD TJ JR, WILCOX KJ, SEGAL NL, RICH S. PersonalitySimilarity In Twins Reared Apart And Together. J Pers Soc Psychol 1988; 54:1031-1039.

TSUANG MT. Genes, Environment and Mental Health Wellness. Am J Psychiatry 2000; 157:489-491.

TSUANG MT. Schizophrenia:Genes And Environment. Biol Psychiat 2000; 47:210-220.

VINCENT JB, KALSI G, KLEMPAN T, TATUCH Y, SHERRINGTON R P, BRESCHEL T, MCINNIS M, PETURSSON H, GURLING HM, GOTTESMAN II, TORREY EF, PETRONIS A, KENNEDY JL. No Evidence Of Expansion Of CAG Or GAA Repeats In Schizophrenic Families With Twins. Hum Genet 1998; 103:41-47.

WILSON EO. On Human Nature. Publ Harvard U Press, Cambridge MA 1979.

About Jewish Genes

BENERECETTI S, SEMINO O, PASSARINO G, TORRONI A, BRDICKA R, FELLOUS, M. MODIANO G. The Common, Near Eastern Origin Of Ashkenazi And Sephardi Jews Supported By Y-Chromosome Similarity. Ann Hum Genet 1993; 57 [Pt 1] 55-64.

HAMMER MF, REDD AJ, WOOD ET, BONNER MR, JARJANAZIE H, KARAFET T, SANTACHIARA-BENERECETTI S, OPPENHEIM A, JOBLING MA, JENKINS T, OSTRER H, BONNE-TAMIR B. Jewish And Middle Easter Non-Jewish Populations Share A Common Pool Of Y-Chromosome Biallelic Haplotypes. Proc Natl Acad Sci USA 2000;97:6769-6774.

LUCOTTE G, DAVID F. Y-Chromosome-Specific Haplotypes Of Jews Detected By Probes 4f A And 49a. Human Biol 1992;64:757-761.

LUCOTTE G, DAVID F, BERRICHE S. Haplotype VIII of the Y-Chromosome is the Ancestral Haplotype in Jews. Hum Biol 1996; 68:467-471.

LUCOTTE G, DAVID F, BERRICHE S. Y-Chromosome-Specific Haplotype Diversity In Ashkenazic And Sephardic Jews. Hum Biol 1996; 68:467-471.

LUCOTTE G, SMETS P, RUFFIE J. Haplotype VIII Of The Y Chromosome Is The Ancestral Haplotype In Jews. J. Hum Biol 1993; 65:835-840.

NEBEL A, FILON D, BRINKMANN B, MAJUMBER PP, FAERMAN M, OPPENHEIM A. The Y Chromosome Pool Of Jews As Part Of The Genetic Landscape Of The Middle East. Am J Hum Genet 2001; 69; 1095-1112.

NEBEL A. FILON D, WEISS DA, WEALE M, FAERMAN M, OPPENHEIM A. THOMAS MG. High Resolution Y Chromosome Haplotypes Of Israeli And Palestinian Arabs Reveal Geographic Substructure And Substantial Overlap With Haplotypes Of Jews. Human Genet 2000; 107:630-641.

SKORECKI K, SELIG S, BLAZER S, BRADMAN R, BRADMAN N, WABURTON PJ, ISMAJLOWICZ M, HAMMER MF. Y Chromosomes Of Jewish Priests. Nature 1997;385:32.

SPURDLE AB, JENKINS T. The Origin Of The Lemba "Black Jews" Of Southern Africa: Evidence From P12f2 And Other Y-Chromosome Markers. Am J Hum Genet 1996; 59:1126-1133.

THOMAS MG, PARFITT T, WEISS DA, SKORECKI K, WILSON JF, ROUX ML, BRADMAN N, GOLDSTEIN DB, Y Chromosomes Travelling South: The Cohen Modal Haplotype And The origins Of The Lemba- "The Black Jews of Southern Africa". AmJ Hum Gnegt 2000;66:674-686.

THOMAS MG, SKORECKI K, BEN-AMI H, PARFITT T, BRADMAN N, GOLDSTEIN DB. Origins Of Old Testament Priests. Nature 1998; 394:138-140.

WWW.MAZORNET.COM/GENETICS/MACHADO. ASP. The Mazor Guide to Jewish Genetic Diseases.

Freud and Schreber

FREUD S. Psychoanalytic Notes on an Autobiographical Account of a Case of Paranoia (Dementia Paranoides) published in1911. [VOL 12 OF The Standard Edition Of The Complete Psychological Works Of Sigmund Freud; Transl Strachey J]. Publ Hogarth Press, London 1958.

SCHREBER DP. [Transl Macalpine,I And Hunter RA]. Memoirs Of My Mental Illness. Publ Dawson Ltd, London 1955

Belief in the Christian Era

DURANT W.—Age Of Faith ; MJF Books, NY 1950.

FEBVRE L. [Transl Gottlieb B]. The Problem Of Unbelief In The16[th] Century. Harvard U Press, Cambridge MA 1982.

HUIZINGA J.[Transl Payton RJ And Mammitzsch U]. The Autumn Of The Middle Ages. Publ U Chicago Press, Chicago 1996.

MANCHESTER W. A World Lit Only By Fire; Publ Little Brown, Boston, 1993.

MARIUS R. Martin Luther-The Christian Between God And Death. Harvard U Press, Cambridge MA 1999.

ORIGEN. [Transl Daly RJ] Treatise On The Passover. Publ Paulist Press, NY 1992.

SAINT AUGUSTINE.[Ed Bourke VJ]. The City Of God. Publ Doubleday, NY 1958.

SAINT AUGUSTINE. Confessions [Transl Pine-Coffin RS]. Publ Penguin Books, London 1961.

SOBEL D. Gallileo's Daughter. Publ Walker & Co, NY 1999.